“The greatest step forward in structural un... ...d practice in the last 200 years”

...ineers, 1983

THE BRITA... ...DGE 1845-1850

A sesquicentennial celebration

Cover: Conference of Engineers for Britannia Bridge by John Lucas

Key:

1	*Admiral Moorsom*	5	*GP Bidder*	9	*Charles H Wild*
2	*Latimer Clark*	6	*Robert Stephenson*	10	*Alexander Ross*
3	*Edwin Clark*	7	*Mr Hemmingway*	11	*Joseph Locke*
4	*Frank Forster*	8	*Captain Claxton*	12	*IK Brunel*

Preface

Without the encouragement of Mike Chrimes, Head Librarian of the Institution of Civil Engineers, this account of the great tubular bridge at Menai would never come to have been written. Without the unstinting help of the Archivist at Great George Street, Carol Morgan, it could not have been written. To them my sincere thanks. So also to all those whose advice and support has contributed to what follows. I must lay personal claim to the errors and omissions

Recorded history has been kind to the memory of Robert Stephenson (1803-1859) but less so to Sir William Fairbairn (1789-1874), a man of at least equal stature but whose career brought him less into the public eye, and who possessed expertise in different fields At Menai, if only briefly, their paths crossed and the outcome was one of the Victorian world's finest bridges.

Stephenson may be said to have been born with a pewter spoon in his mouth, for, at the time of his birth, his father, George, was already prospering, and could afford to give his son a sound education. In later years the firm of George Stephenson and Son provided a springboard for Robert of which he took full advantage to launch his remarkable career.

Fairbairn, on the other hand, was the son of a farm worker in the Scottish Borders, had little schooling, and raised himself by his own prodigious efforts to become a major industrialist, and a leading authority on iron, its production and working, as well as being widely recognised by his contemporaries as an experimental scientist and Fellow of the Royal Society.

The third character to play a leading part in the Britannia Bridge story was Edwin Clark (1814-1894). In his youth he had been something of a lad about town, even a ne'er do well. After an uncompleted course at Cambridge, studying mathematics, he turned to teaching before he achieved his meteoric rise from obscurity under Robert Stephenson's wing.

It is hoped in the story which follows it becomes clear that this was a team effort and equal honours are due to each man in his own right. The Britannia was not just one man's bridge.

Samuel Smiles and the many historians who trod in his footsteps have much to answer for.

John Rapley,
Honiton, 1999

Contents

ISBN 0 7277 2847 4

Designed by Rob Norridge/Norridge Walker Design Associates

Printed by Thanet Press Limited, Margate

Britannia Bridge by G Hawkins

Introduction

Mr Stephenson putting in the last rivet

On 5 March 1850 Robert Stephenson drove the first train to pass from the Welsh mainland to the island of Anglesey. At noon he, with the assistance of Mare, the contractor for seven of the eight tubes, placed the last of more than half a million rivets. In his forty seventh year Stephenson stood at the peak of his career, and with the opening of the first Britannia tube a milepost in engineering history had been passed.

The Britannia Bridge, and its smaller partner at Conwy, were like nothing that the world had seen before. In 1845, when the bridges were first conceived, the very wrought iron of which they were to be built was almost an unknown quantity. Its properties as a structural material had been little studied, and only the ship builders had used it on any scale and possessed any understanding of its large scale working.

Stephenson's concept of using a large iron tube as a beam or girder appears to have been original despite rival claimants. What Stephenson failed to appreciate was that a circular or elliptical tube is not the shape best suited to act as a beam, although natural forms such as plant stems were often cited as examples. Nature has perfected plants to resist forces from any side, but a beam is mainly loaded in one direction, and a far more efficient distribution of material is possible. The running of a train through a tube was a logical consequence if, for strength, the tube had to be made sufficiently large.

The method of erection was novel. Few rigid bridges had been built without the benefit of some form of scaffolding, which the Admiralty would not countenance at Menai. The system finally adopted, after much cudgelling of brains, had never before been practised. Prefabrication of the tubes on land, followed by floating out into position and lifting by hydraulic power up to their final resting place, was so new as to invite dire warnings from even the best informed critics.

The introduction of the principle of structural continuity, widely known but little understood, for the first time on such a scale ensured that the Britannia Bridge, designed for the loads of the 1850s, would carry the much greater loads of a century later.

Robert Stephenson was a natural conservative in matters of civil engineering, preferring to adopt tried and tested methods. At Menai he had left himself with few options. There were no precedents to guide him, but having undertaken to bridge the Straits there was no going back or his hard won reputation would lie in ruins. There was perhaps one man in Britain uniquely fitted to offer guidance and practical help, and that man was the industrialist, ironmaster, ship builder, and experimental scientist, William Fairbairn of Manchester.

The early history of the Holyhead route to Ireland

From early times Holyhead, on Holy Island at the north western tip of Anglesey, has provided a convenient point for crossing the Irish Sea to various small harbours in the vicinity of Dublin. The abolition of the Irish Parliament after the Act of Union of 1800, with the resultant centralisation of so much administrative power in London, made improved communications essential, and between 1815 and 1818 Thomas Telford (1757-1834) drove a new coach road through Snowdonia to Bangor, and also effected major improvements elsewhere. One problem remained to be overcome, and that was the ferry crossing the narrow but often turbulent Menai Straits. By 1826 this too had been conquered by the completion of Telford's magnificent Menai Suspension Bridge, with its previously unheard of span of 580ft.

With the coming of the railways various proposals emerged for new ports on the Welsh mainland, with the intention of avoiding the Menai crossing. However, the Admiralty had already chosen Holyhead for major development as a "harbour of refuge", and the arrival at Chester of the railway from Crewe in the spring of 1840, offering a direct link to London, brought the Holyhead route to the fore once again. Two giants of the Railway Age battled it out with opposing schemes. George Stephenson (1781-1848) favoured the route originally proposed by Francis Giles (1787-1847), which followed the North Wales coast from Chester, and he proposed to cross the Menai Straits by laying a line over Telford's bridge, using horses to draw the carriages across. C.B. Vignoles (1793-1875) preferred an alternative route from Shrewsbury, passing through Bala, to a new harbour at Porthdynllaen on the mainland west of Pwllheli.

Various reports were prepared, and there was a keen political debate, but finally a report by James Walker (1781-1862) in October 1843 came down firmly in favour of the Stephenson line, but with a new rail bridge over the Menai Straits. The Admiralty too hoped

Telford's Menai Suspension Bridge

Burdon and Wilson's Sunderland bridge under construction

that the Railway Company would contribute substantially to the £700,000 which the new harbour works would cost. The first Chester & Holyhead Railway (C&HR) Bill was read in March 1844 and received Royal Assent only four months later, with Robert Stephenson as Chief Engineer. However, details of a short section including the bridge were left until the next Session in 1845. Crossing the Straits at a point where the Britannia Rock outcropped at midstream was proposed by Stephenson, and accepted as the only viable option, whatever the final design of the bridge itself might be. The total width of the channel at that point is about 1,100ft.

Early proposals for the crossing

Stephenson's first choice was for a double arch cast iron bridge, with spans of 360ft. While cast iron arches greater than this had been proposed elsewhere, the longest actually constructed up to that time was only 240ft in span. Stephenson also had to face the fact that the Admiralty would permit no obstruction of the navigation. Furthermore, the Admiralty, after receiving reports from Sir John Rennie (1794-1874), Captain Vidal RN, and J.M. Rendel

Stephenson's cast iron arch proposal for the Britannia Bridge

(1799-1856) firmly vetoed any suggestion of an arched bridge, and demanded a minimum clear span of 450ft with 105ft headroom at all points over the navigational channels.

Stephenson had already amended his proposal for the erection of a two arch bridge to constructing one on similar principles to Burdon and Wilson's span of 236ft at Sunderland, built in 1796 without centring. But whereas Wilson cantilevered out the two halves of his arch with ties back to the bank until the centre joint could be made, it was impracticable to build a two arch bridge in this fashion for lack of any anchorage at the midstrean pier. Stephenson proposed to overcome this by the use of balanced double cantilevers at each pier, anticipating by forty years the method used so successfully at the Forth Bridge. At the centre of the spans, where the opposing cantilevers met, his intention was to leave expansion gaps and in consequence, unlike Burdon's bridge, true arches would not have been created. The demands of the Admiralty put an end to these proposals also and they were abandoned but only after drawings had been prepared.

Stephenson was well aware of the inadequacies of suspension bridges when carrying concentrated rolling loads. Captain Samuel Brown's (1776-1852) bridge at Stockton on Tees had proved totally unsuitable for rail traffic. It had to be propped up, and later replaced by Stephenson himself with a trussed cast iron girder bridge of a type which, in its turn, was to fail at Chester in 1847. Stephenson recognised that the weakness of a conventional suspension bridge lay, to a considerable extent, in the flexibility of the deck, and surmised that by making the deck sufficiently stiff this problem could be satisfactorily overcome.

What he failed to appreciate was that the greater the vertical stiffness of the deck, the greater the bending forces at the points of attachment to the catenary chain, and this was later to be a deciding factor in the final decision to forego chains as a permanent form of support for the tubes. Fairbairn considered any attempt to interface a rigid deck with a flexible support to be positively dangerous.

Iron bridges before 1845

Abraham Darby's cast iron arch over the Severn at Coalbrookdale with a span of 100ft was completed in 1779, the first major bridge to be built entirely of iron, and it was followed some years later by Telford's Buildwas bridge of 1796, a few miles upstream, having a span of 130ft. In design, Buildwas was the forerunner of the cast iron arch rather than Darby's pioneering concept which was based on the mode of construction of timber bridges and was designed by a carpenter. By 1845 the cast iron arch had become well established, and it was only to be expected that Stephenson would turn first to such a well tried and reliable form of construction. Unfortunately he had reckoned without Captain Alexander Vidal and the Admiralty.

The simple cast iron beam, immensely popular for small spans, had reached its practical limit of around 50ft. The Midland & Great Western Railway of Ireland had built a bridge of this size in 1845, with 52ft girders, each cast in one piece. Trussed cast iron girders, which were within a short time to be condemned as structurally objectionable, had not yet been built for a span as great as 100ft. Clearly any form of cast iron beam was impracticable at Menai.

A brief note on the differences in the manufacture and properties of cast iron and wrought iron may not be out of place. In both cases the raw iron ore was smelted in a blast furnace, and the molten iron run into moulds in the floor resembling a sow with piglets, hence the name pig-iron. With or without refining, pig-iron could be melted and poured into moulds which allowed the formation of intricate shapes in one operation. Unfortunately such castings tended to be brittle and unreliable in tension but possessed great resistance to compression. Fairbairn was right to describe cast iron as a treacherous material when used in tension.

Wrought iron was produced by refining pig-iron in a puddling furnace, from which it was extracted in blooms of several hundredweight, and hammered into a rough shape which could be further formed in a rolling mill to give various shapes such as round and flat bars, and angle and tee irons. This quality was known as merchant bar. By a further process of

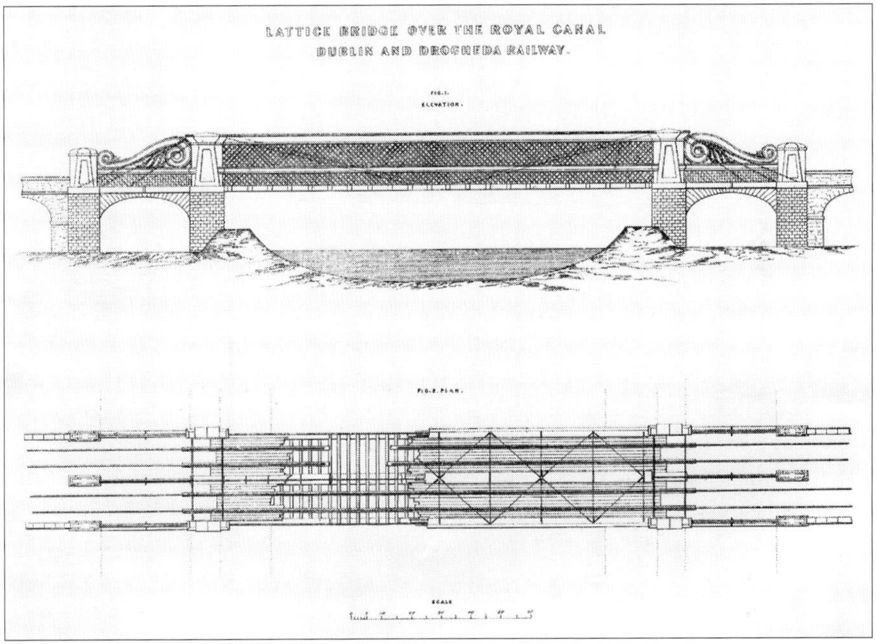

cutting merchant bar into short lengths laid together, reheating and hammering them into a solid mass before re-rolling, a better quality iron was obtained, more suitable for plates. This was known as "best" iron. Repeating the progress once or twice again produced "best best" and "best best best" iron. The last was mainly used for boiler plates, but at Britannia it was specified for the bottoms of the tubes where they came under the greatest tensile stress.

Wrought iron was still in its infancy as a bridge building material in 1845. It was expensive, very roughly twice the price of cast iron, and complex shapes could only be formed by fabrication. But it had great tensile strength, and resistance to shock loads. It lent itself well to hammer welding. The ship builders had long recognised its virtues, but the bridge engineers were slow to appreciate its potential.

The most significant development took place in Ireland soon after 1840. Sir John Macneill (*c*1793-1880), assisted by James Thomson, had built an experimental road bridge of 80ft span to cross the Dublin & Drogheda Railway, then under construction. This was based on Ithiel Town's American patents of 1820 and 1835, which were intended for timber construction. The sides of Town's bridge were trussed by the use of a lattice or trellis of small timbers crossing approximately at right angles. Macneill adopted this idea but used flat iron bars with iron top and bottom chords. He designed two bridges of 140ft span to cross the Royal Canal near Dublin and a city street. Unfortunately this pioneering design was not without its problems but, according to John G. James, they were the first true iron lattice bridges.

Elsewhere in Europe timber remained the predominant material, although Russia and Sweden had well developed wrought iron industries producing material of excellent quality. However, like America, they too had almost boundless supplies of cheap timber, and iron bridges were consequently slow to develop, but in Russia much development work was done on Howe trusses, and it was there that the forces in open webs were first evaluated. Nowhere was there anything remotely approaching in scale the conditions Stephenson had to overcome at Menai. The only existing bridges of comparable span were all of the suspension type, and their limitations were manifest to all.

Sources of information

The story of the Britannia and Conwy tubular bridges has been told many times in the past century and a half, but apart from a few authors alive at the time, none of whom were actually engaged in the project, the primary source of information has been two books. The more modest volume was published by Fairbairn in 1849 after his connection with the bridges had ceased. The following year Edwin Clark's more imposing work caught the public eye. Written under the supervision of Robert Stephenson, it combined an account of the bridges with a textbook on structural engineering. The mathematical design sections were written by William Pole (1814-1900) at Clark's request. Fortunately for students of engineering history both these volumes, and much other invaluable reference material including Clark's working diary, has been carefully preserved in the archives of the Institution of Civil Engineers.

It is impossible to reconcile the two accounts completely, but from its pedestrian style and self-congratulatory tone Clark's book lends credence to Fairbairn's acid comment "*It seems to have been got up for the praise of Stephenson and his school, - claiming the merit of a great discovery, to which they have not a shadow of a title, beyond a very crude idea of a wrought-iron suspended cylinder, which I afterwards proved to be useless*".

Although Fairbairn's name occurs in a number of references in Clark's book, they play down the importance of Fairbairn's contribution, and attribute to Stephenson decisions which only confirmed those already made, or were an acceptance of advice given. It may fairly be argued that without a man of William Fairbairn's knowledge and experience the tubular concept would have been stillborn. Fairbairn had a justifiable complaint, especially since he himself had paid a generous tribute to Clark in his own book the year before.

Neither account is entirely trustworthy; both authors had axes to grind, and set about it with a will. In Fairbairn's case, he was anxious to establish the crucial part that he had played in all the development and design work, and also the construction up until completion of the first Conwy tube. Clark, writing on behalf of Stephenson, attempted, unjustifiably but with a considerable degree of success, to claim all the credit for his master and himself. Factual statements, where supported by clear evidence, can be accepted with some confidence, but the many unsupported assertions call for careful scrutiny and some considerable degree of scepticism. This applies also to later accounts, which have in the main been based on Clark's work.

While Clark's book is the only full account available, it is written with the benefit of hindsight and in consequence there are deliberate omissions and ambiguities. One which will be dealt with later is the construction of the Britannia Bridge as a pair of continuous beams. The idea of using the shorter land spans as a counterpoise to the main tubes obviously originated from an earlier idea of using balanced cantilevers in order to obviate the need for scaffolding, but beyond that Clark is unfortunately silent as to who suggested and advised on the practical application of continuity. The adoption of this mode of construction was crucial to the long term future of the bridge, and Clark himself would seem to have advocated it. Fairbairn was far from fully convinced of its benefits.

William Fairbairn

Conceiving the idea

An innovative solution

Stephenson considered but rejected several alternative schemes for which there was no precedent, before finally conceiving the idea of an iron tube, hung from chains, and sufficiently large to allow the passage of a train. As so often happens, others were later to lay claim to having originated the idea. Among them was his colleague, J.M. Rendel, who had supported the idea before the Parliamentary Committee, and a John de la Haye of Liverpool, A Frenchman, Dr. Jules Guyot and others also staked their claims. To his credit William Fairbairn, even after his dispute with Stephenson, unreservedly acknowledged the latter's claim to priority.

It was at this juncture that Stephenson, who had consulted widely as to the practicability of his concept, invited the views of William Fairbairn when the latter was calling on his old friend from Newcastle days, George Stephenson, at the firm's office at No.24 Great George Street in April, 1845. George had of course retired from active participation by this date, but continued to take a great interest in his son's projects. At this stage neither drawings nor calculations existed, but Fairbairn had been constructing large iron ships since 1831, and appreciated that the suggested tubular bridge deck had much in common with a ship's hull, which acted in principle like a beam. As a result of this chance meeting, agreement was quickly reached for a series of experiments to be carried out at Fairbairn's shipyard at Millwall, in order to arrive at the most suitable shape of tube to be adopted, and to establish the feasibility of such a project, for they were entering uncharted waters.

The first series of experiments

Within a month Stephenson was called upon to give evidence to the Parliamentary Committee considering the second Chester and Holyhead Railway Bill, which included the bridge at Menai. Stephenson outlined his proposal to erect two platforms, each supported by three iron chains, to span the distance from the Britannia pier to the two shore piers, a matter of 450ft. On these platforms he planned to build two pairs of elliptical tubes, one for each track, with an internal height of 25ft and a width of 15ft. The tubes were to be fabricated from iron plates, riveted together, and stiffened with angle irons. The top and bottom were to be of 7/8in plates, and the sides of 1/2in material. A timber or iron frame internally would support the permanent way. The weight of each tube was estimated at 450 tons. Drawings were prepared and calculations made by two assistants, G. Berkley and W.P. Marshall, and these were considered by Stephenson to establish the practicability of the concept.

When presenting his evidence Stephenson was asked if his calculations had been submitted to many other engineers. He replied "*I have made them in conjunction with Mr Fairbairn of Manchester, whose experience is greater than any other man's in England.*" When further asked if the chains would form a permanent part of the finished bridge, his answer was non-committal "I think it probable that they would be left as a prevention, but yet I find that the tube itself would be quite sufficient to support any ordinary railway-train." The proposal was supported by J.M. Rendel, and General Pasley RE, Chief Inspector of Railways for the Railway Commissioners. The Act received the Royal Assent at the end of June, 1845.

Payment of Fairbairn's expenses was approved by the Chester & Holyhead Directors, a sum of £3,000 being agreed. He was given a free hand to press ahead with his experiments entirely at his own discretion. Stephenson was content to leave the matter in Fairbairn's hands and managed only two short visits to the site while work was in progress. There was also a visit from several of the Directors and the Company Secretary, George King. Otherwise Fairbairn was left to his own devices, Stephenson and his staff being fully occupied with planning and Parliamentary

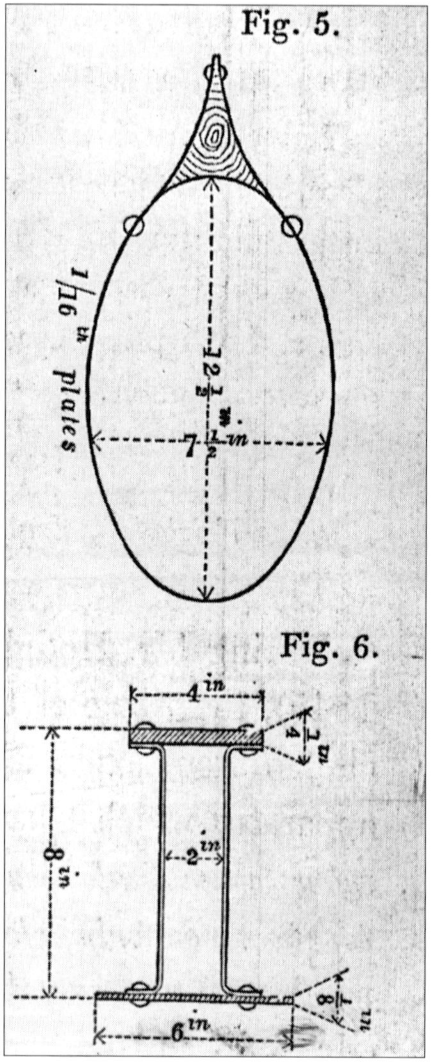

Experimental tube of elliptical section

work for some thirty four railway schemes at home and others on the Continent.

For his own part, Fairbairn had a substantial business to run in Manchester, employing at times as many as two thousand men, and in between his frequent visits to Millwall much of the experimental work was delegated to his three foremen, Henry Ross, James Graham, and Robert Murray, supervised by his son Thomas, a partner in the business, and manager of the shipyard. The reports on the first series of experiments in Clark's book were later compiled from Fairbairn's detailed notes, since Clark did not enter Stephenson's service until several months after the initial series of experiments had been satisfactorily completed. This he failed to make clear in his book.

As the experiments progressed Fairbairn realised that the work would call for an experienced assistant, with a deeper knowledge of mathematics than he himself possessed, and he sought permission from Stephenson to engage his old friend and fellow experimenter, Eaton Hodgkinson FRS (1789-1861). The aim was to use Hodgkinson's mathematical expertise to develop a general formula for assessing the strength of iron tubes. On receipt of Stephenson's approval Hodgkinson first visited the Millwall site on the 19th. of September 1845, but by then the initial experiments were almost completed. Hodgkinson had to delay a start until he had finished work already in hand, and it was not until some days later that he was free to commence at Millwall. As theory would suggest, it had very quickly became obvious that rectangular tubes were superior in strength, although with all types the tops were generally the first parts to fail by buckling. Fairbairn was later criticised for wasting time on round and elliptical tubes, but this was done at Stephenson's behest since he strongly favoured the elliptical shape.

It was then appreciated, apparently for the first time, that wrought iron lacked the great resistance to compression exhibited by cast iron. Not only was it found that the top of a girder required a greater cross section of iron, but that the distribution of the metal was also critical. Formation of the top into a series of circular or rectangular cells was demonstrated to be the most effective solution, and with this discovery Fairbairn was able to write "*It is from this period that I date the disappearance of almost every difficulty respecting the construction and ultimate formation of the Britannia and Conway (sic) tubes.*" As early as 20 September 1845 Fairbairn had sent Stephenson a sketch of a tube with square cells top and bottom, almost exactly resembling those actually built. This was before Hodgkinson had even started work at Millwall.

A progress report to the directors

At the beginning of 1846, under considerable pressure from the Directors, Stephenson was called upon to furnish a progress report, to which he added independent reports from Fairbairn and Hodgkinson. Stephenson's own report dated 9 February 1846 stated that his belief in the efficacy of a wrought iron tube had been vindicated by the results of the experiments, but he conceded, albeit reluctantly, the superiority of a rectangular form of tube with extra metal in the top. He recognised that the use of chains in the conventional manner would be totally unsuitable, and stated that if, after further experiments, the use of chains could not be avoided, some new method of attachment must be devised to maintain the integrity of the tubes. In conclusion, he reported that work on the masonry piers was due to commence early in April.

Hodgkinson's report was mainly technical, but he was clearly unhappy with, as he saw it, the incompleteness of the experiments. He concluded by saying "*If it be determined to erect a bridge of tubes, I would beg to recommend that suspension chains be employed as an auxiliary ...*"

Fairbairn had no such qualms. After detailing his experiments he cane down firmly in favour of a rectangular tube with a cellular top, but recommended further experiments on a larger scale model. He stated his opinion that a tubular bridge could not only be constructed to carry railway traffic, but that "*it should be a huge sheet-iron hollow girder, of sufficient strength and stiffness to sustain those weights ...and you may strip off the chains...*" He never wavered from these views, while Stephenson preferred to keep his options open. In any case, at that stage chains would be needed for the planned method of erecting the tubes and, if not required permanently, could to be sold for use elsewhere.

Hodgkinson became increasingly restless. He had been sorely disappointed to find that he was regarded merely as a technical assistant rather than director of the experimental work. Now he feared that the engineers were rushing ahead recklessly, without adequate data, as a result of commercial pressures, and he was by nature cautious and pedantic in his approach. Fairbairn, on the other hand, never lacked confidence in his own abilities, far more so than Stephenson who suffered his moments of doubt. The pace at which the two engineers were prepared to push forward into unknown territory was remarkable by any standards, and almost incredible today. But Stephenson was anxious to

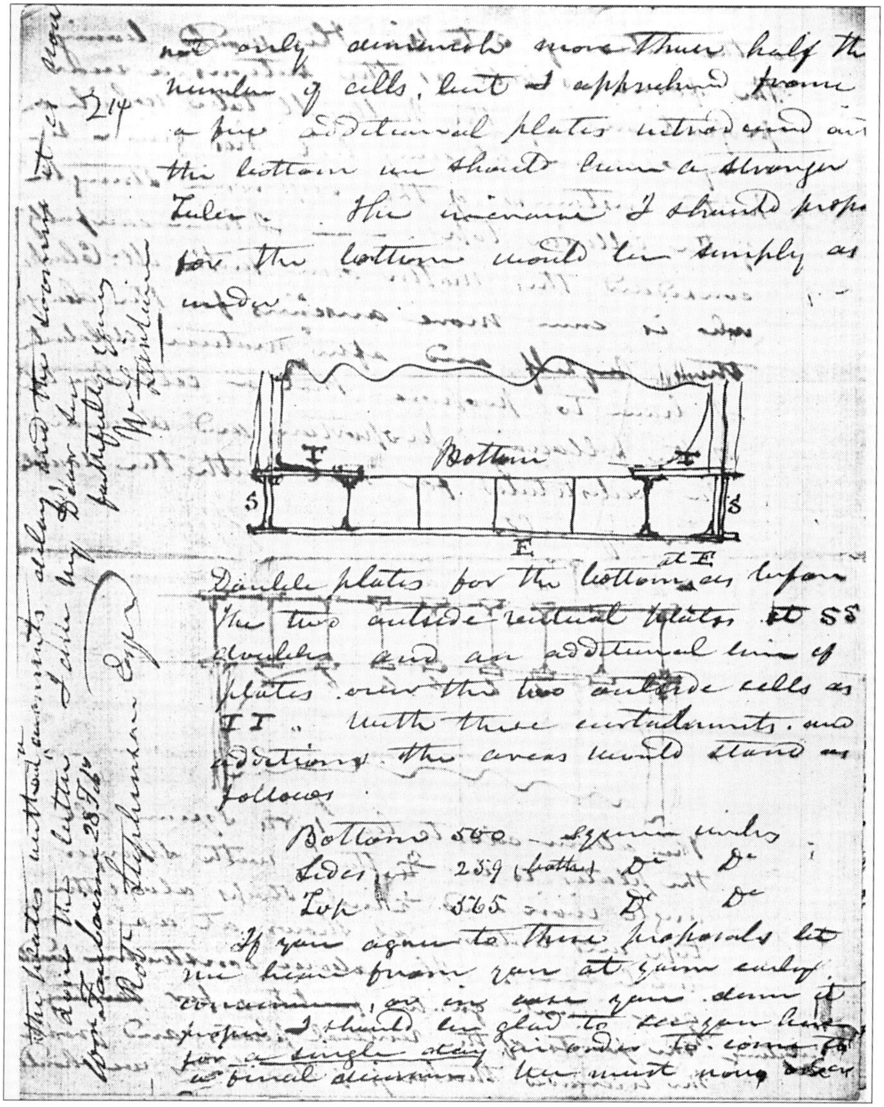

not only diminish more than half the number of cells, but I apprehend from a few additional plates introduced in the bottom we should have a stronger Tube. The increase I should propose for the bottom would be simply as under

Bottom

Double plates for the bottom as before at E. The two outside vertical plates at SS doubled, and an additional line of plates over the two outside cells as II. With these curtailments and additions the areas would stand as follows

Bottom 560 square inches
Sides 259 Do Do
Top 565 Do Do

If you agree to these proposals let me hear from you at your early convenience, or in case you deem it proper, I should be glad to see you here for a single day in order to come to a final decision. We must now …

Letter from William Fairbairn to Robert Stephenson

placate the Directors, who were becoming impatient under pressure from the shareholders who in their turn were complaining loudly of lack of progress.

Erection of the masonry towers for the Britannia Bridge

Hodgkinson made clear his feelings in a letter to Stephenson early in March, 1846, to the displeasure of Fairbairn, who complained that Hodgkinson viewed the experiments as an abstract scientific investigation and was oblivious to time and commercial considerations. To smooth things over, Stephenson agreed to Hodgkinson continuing his experimental work, leaving Fairbairn free to develop his ideas and to carry out tests on a scaled down version of the actual tubes as proposed. Even at this early stage the overall dimensions of the full size tubes had been provisionally agreed in order to allow work to proceed on the foundations and piers.

Fairbairn's patent for box girders

Early in 1846 Fairbairn had recommended to Stephenson adoption of wrought iron box girders for the proposed C&HR bridge over the Dee at Chester, but without success. A trussed cast iron girder bridge was chosen instead, and this was to prove an unfortunate error of judgment. With the proven superiority of the rectangular tube, Fairbairn could envisage future prospects in using smaller tubes for short to medium spans, and with Stephenson's reluctant approval he took out a Patent, No.11,401, dated 8 October 1846. Stephenson shared the costs, but preferred not to have his name involved. He had a personal objection to patents, considering them to stand in the way of progress. Not so Fairbairn, a true Victorian entrepreneur.

Earlier trials with single web plate girders, imitating the shape of Eaton Hodgkinson's cast iron girders, had failed due to buckling and twisting under load as a result of inadequate top flanges, and their use was limited mainly to deck girders for ships where the deck plates which provided stability. In later years Fairbairn came to appreciate the greater economy of a properly proportioned single web girder, and manufactured them in considerable quantities. In practice within a few years the box section was to become obsolete for smaller spans.

From April 1846 Fairbairn's responsibilities became ever more onerous. He was no longer a contractor carrying out a series of experiments, but assuming major responsibilities which Stephenson had neither the time nor the expertise to undertake himself. To regularise the position, at a meeting of the Directors in mid-May, Fairbairn was formally appointed, on Stephenson's recommendation, "*to superintend the construction and erection of the Britannia and Conway (sic) Bridges, in conjunction with Mr.Stephenson*" at a salary of £1,250 per annum. He not unnaturally interpreted this as making him Joint Engineer for the bridges, and was confirmed in his views by a later letter from Stephenson in August 1846 - "*I consider you as acting with me in every department of the proceedings*". Sadly, he was to find that Stephenson had other ideas when it came to sharing the credit, in due course claiming it all for himself, and avoiding any but the barest acknowledgement of Fairbairn's vital contribution.

Francis Thompson (1808-1895), the architect responsible for major buildings on the C&HR, was briefed, and requested to prepare contract drawings in order that the work on the piers could go out to tender. The towers were to include provision for chains. Frank

Forster (1800-1852) was appointed Resident Engineer for the Britannia piers, and Alexander M. Ross (*c*1805-1862) was to supervise the bridge at Conwy. Thompson lost little time, Nowell, Hemingway & Pearson were awarded the contract, and the foundation stone of the Britannia pier was laid by Forster on 10 April 1846. Ross laid the Conwy foundation stone a month later, and John Evans, a very experienced contractor from Cambridge, was awarded the contract for the fabrication and erection of the Conwy tubes as well as the masonry of the foundations and piers.

The tubular bridge at Conwy

As at Menai the bridge at Conwy had been planned as twin tubes, but only a single span of 400ft was required. It had originally been proposed by the C&HR Board that both bridges should proceed simultaneously, much to Stephenson's displeasure, but in the event the first Conwy tube was completed earlier, and provided a useful testing ground before construction of the Britannia tubes had progressed too far.

Fairbairn gave instructions for a model tube to be constructed at Millwall, 1/6 of the size of the proposed Britannia tubes. In earlier tests he had tried circular cells to strengthen the top against buckling. Now he returned to the idea of rectangular cells as being easier to fabricate and maintain, and almost equally effective. During the first week in June 1846 the model tube was completed and given its initial test. After each failure the tube was progressively strengthened to eliminate weak parts, and the final experiment completed on 11 September. By this time Fairbairn had accumulated enough data to derive a formula to compute the strength of tubes of similar proportions but of any length, and in this he had the assistance of Dr Thomas Tate of Battersea, Hodgkinson being still engaged on an apparently fruitless series of experiments in Fairbairn's works at Manchester. The way was clear to scale up to the spans required at Britannia and Conwy.

To conclude the development work Stephenson was invited to a demonstration at Millwall in the second week of October 1846, when the final experiment was made on the model tube. For advice on the method of scaling up from the model, Fairbairn turned to Sir George Biddell Airy FRS (1801-1892), the Astronomer Royal, an exceptionally gifted mathematician with an interest in solving practical engineering problems. However, as Airy was to point out, Galileo had discovered the answer back in the sixteenth century when he

Conwy tubular bridge

propounded the "square-cube law".

Airy's advice was simple. If the dimensions of the model - i.e. length, breadth and depth - are increased by a factor *x*, the weight will increase by *x* cubed. Unfortunately such a simplistic approach is full of pitfalls. To quote Professor Gordon of Reading, a modern authority, "*The strength of any structure which is liable to fail because the material breaks cannot be predicted from models or by scaling up from previous experience.*" Fairbairn was taking a leap into the dark, and would without a doubt have informally consulted two of his eminent friends, Henry Moseley MA FRS (1801-1872) and Professor Willis BA FRS (1800-1875).

Edwin Clark joins Stephenson's staff

During the early Spring of 1846 there had been another development which was to have a considerable bearing on the future of the bridges, and of Fairbairn personally. Some time in March, after the first report to the Directors, a young man, 32 years of age, called at Stephenson's outer office with an introduction, and would take no discouragement by the Chief Clerk, John Sanderson, Stephenson's brother-in-law. Personable, and with a silver tongue, Edwin Clark so ingratiated himself with the great man that he speedily found himself installed in an office as Stephenson's personal assistant to supervise and report on the tube experiments. For one without any previous engineering education except an uncompleted course in mathematics, Clark's later promotion to Resident Engineer on the Welsh bridges was, perhaps, the most remarkable appointment in the history of engineering. At the same time it demonstrated Stephenson's considerable capacity for delegation, and for choosing subordinates on whose skill and loyalty he could rely while at the same allowing them freedom to develop their talents.

How the Britannia Bridge was to be erected still remained unresolved. The first proposal - to assemble the tubes on a platform suspended from chains - was clearly impracticable. Stephenson then suggested that the tubes should be constructed on the land approaches, and rolled out over the platform which had been pre-loaded with weights. As the tube advanced the weights would be withdrawn, thus maintaining a constant level. Fairbairn had suggested a double cantilever system, working out from the piers, and much as had been proposed for the arches, but on mature reflection he felt this to be too dangerous.

It is claimed by Clark that, while waiting on Crewe Station, he had watched a large water tank being raised into position with screw-jacks, and he suggested this method for raising the bridge tubes. John Evans, the contractor at Conwy, took up this idea with enthusiasm and he is credited with the idea of building the tubes on staging at the waterside and floating them into position. He had the courage of his convictions, and undertook full responsibility for the floating and lifting of the Conwy tubes. At Menai it was decided that the land spans should be assembled on staging in their final position, and the main tubes assembled on staging along the shore line. Tides in the Straits run fiercely, and there is only a very short period of slack water at high tide, but no other method seemed feasible. Even so, some argued that the chains might still be needed to assist in the

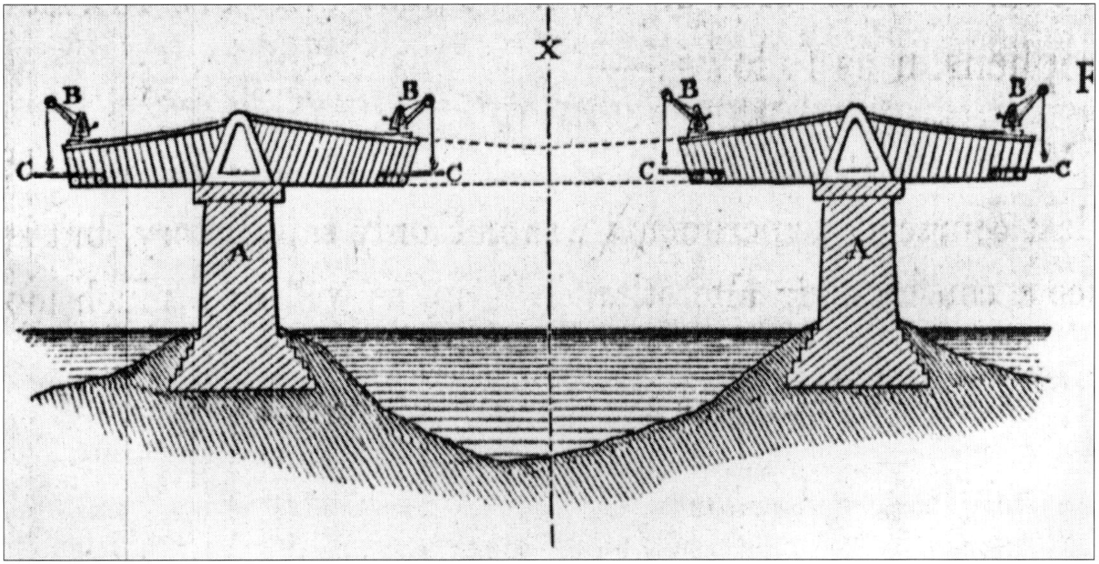

Proposed method of cantilever erection of the Britannia Bridge, not adopted

lifting. Fairbairn strongly disagreed, placing his faith in a system which he devised, using hydraulic jacks and lifting chains.

It was in mid-July 1846 that Fairbairn put these proposals to Stephenson, who was then deeply involved in Parliamentary work for three railway companies, while at the same time being under intense pressure from his doctor to rest. The strain of his enormous workload was starting to undermine a constitution which had never been particularly strong. In spite of some misgivings Stephenson so far accepted the scheme that, a month later, he was urging Fairbairn to proceed with the staging and workshops for the fabrication of the tubes. At the same time Fairbairn was investigating the supply of lifting chains, and also in mid-July his drawing office in Manchester started on the working drawings for the tubes.

Drawings for the workshops had already been put in hand in mid-May, and were completed in a month. Fairbairn wasted no time, and within four months the shores at Menai were hidden under a massive timber staging, 3 acres in area, together with extensive workshops, which included the latest tools and machinery, such as Roberts' Jacquard automatic punching machine, Fairbairn's steam riveters, and rolls for flattening the plates. At Conwy this operation was carried out with 40lb sledgehammers on a cast iron stretching block. With an eye to economy discarded tubes from the Millwall tests were adapted as chimneys for the rivet heating furnaces, forty eight in all.

Disputes over priority for the cellular concept

Relations with Hodgkinson were becoming ever more strained, as the latter was claiming priority for the idea of cellular tops. Matters came to a head at a meeting of the British Association at Southampton, where both he and Fairbairn presented Papers, and Hodgkinson publicly repeated his claim. To add to the confusion, Stephenson, in a letter to Fairbairn in late October, claimed to have developed a cellular construction for a bridge at Ware some years earlier. He wrote "*It was I believe a counterpart of the proposed top of the Britannia Bridge.*" When challenged, Stephenson produced a drawing of a built up wrought iron I beam, bearing no resemblance to the intended cells. This did not prevent Stephenson and Clark from repeating the claim later, although it was admitted that the apocryphal design was never built. Fairbairn's three foremen at Millwall gave written statements supporting his claim to priority, but by now the seeds of dissension had been well and truly sown.

Towards the end of September Stephenson had to go abroad for some six weeks, and when he returned in early November the contract drawings had been completed in Fairbairn's drawing office, under the supervision of his personal assistant, Mr Blair, and had been signed by Fairbairn himself. On Stephenson's return the drawings were presented to him for approval in order to allow the tubes to go out to tender without delay. Regardless of a mere seven months experience and having no training in drawing office work, Clark was later to claim that he had personally completed the drawings.

Stephenson, deeply involved in so many other projects and in indifferent health, was able to devote little time to the bridges, and became much concerned that Fairbairn was stealing his thunder. Tensions between the two men started to show. There must inevitably be a suspicion that Clark, growing increasingly confident and trusted as a go-between by Stephenson, may have used these differences to enhance his own position. To further strengthen this he was able to persuade Stephenson to appoint his brother, Latimer (1822-

1898), as his personal assistant when he himself was made Resident Engineer.

With the design of the tubes finalised Fairbairn turned his attention to the supplies of iron. Provisional orders had already been placed with a number of ironmasters, since the mass production of wrought iron was still at an early stage of development, and a large part of the available output was absorbed by the rapidly expanding ship-building industry. In addition, there were problems over quality, and few works could roll the size of plate needed for the bottom cells. In total some 12,000 tons of good quality iron would be required. By February of 1847 enough iron had been received at Conwy for Evans to start punching plates, but a start on the Britannia tubes was delayed by lack of material. There were problems with quality, and a parcel of ten tons of plates from Coalbrookdale, with whom Fairbairn had negotiated a special price for superior quality, was condemned and returned.

Contracts for the tubes are awarded

The contract for the first Britannia tube was awarded to Garforth of Dukinfield, but it had been expected that Fairbairn's Millwall yard and Ditchburn and Mare of Blackwall would share the fabrication of the remaining seven Menai tubes. Fairbairn's appointment as a salaried engineer now involved a conflict of interest. To avoid this, Thomas Fairbairn negotiated an arrangement with Ditchburn and Mare for them to undertake the whole work at the same price. To the annoyance of the Directors they had not been consulted, and there were nasty rumours that Fairbairn's firm was receiving a good commission. The matter was smoothed over, but at a meeting of the Finance Committee in January 1847 some strong criticisms were voiced concerning the continuing outlay on Hodgkinson's experiments, which appeared to be getting nowhere despite the full time employment of six or seven of Fairbairn's men. Regardless of their differences Hodgkinson had continued to enjoy the facilities of the Manchester works. The cost had now risen to over £6,000, double that originally approved, and Stephenson agreed that it was time to call a halt.

This he seems to have been reluctant to do, for Hodgkinson was urging the use of a cast iron top for the tubes, since his model tests were proving satisfactory. Clark wrote to Stephenson at the end of February 1847, reporting these findings, and Stephenson commended to Fairbairn the use of cast iron tops for box girder bridges, although accepting that the tubular bridges could not be modified at that late stage. It was also at this time that, with Fairbairn's approval, the top cells were reduced to a single row and increased in size to permit access for riveting and painting. Stephenson had some smaller box girder bridges built with cast iron tops, although after constructing four the idea was abandoned.

Lifting machinery for the tubes

While agreement had been reached on floating out the tubes, the machinery for lifting them was still under consideration, and once again it fell to Fairbairn to devise the means. Design work started at the end of January 1847, and Fairbairn was able to report two months later that the drawings had been completed and that enquiries were being made regarding the hydraulic presses required. The proposed scheme involved massive cast iron cross girders carrying the presses and situated near the tops of the piers above the highest point of lifting. In turn the presses had a crosshead carrying chains from which the tubes would be suspended. The presses were to have a stroke of six feet so that the full lift would be carried out in a number of stages, the masonry being built up under the end of the tube as each stage was completed. At Conwy a single press at each end of the tube would suffice, but at Britannia, where a much greater weight was involved, the smaller presses from Conwy would be used as a pair at one end, and a third and proportionately larger press at the other. One of the crosshead castings at Conwy developed cracks while lifting was under way, but fortunately it held. In the light of this potentially dangerous weakness fabricated wrought iron components were later adopted for the heavier loads at Menai.

The chains were of a newly patented type, where the end eyes were rolled in one piece with the link, instead of being welded on as had been the custom previously, resulting in a far more reliable product. Fairbairn approached the patentee, Thomas Howard, of Howard and Ravenhill at the King and Queen Ironworks, Rotherhithe, and an order was placed. Each link was 7in wide and 1in thick, and 6ft between the centres of the eyes. Great precision was necessary so that each link would bear its fair share of the load when coupled in sets.

The pumps and hydraulic presses were supplied by Easton and Amos of Southwark, who specialised in extrusion presses for seamless lead pipes. Each pump was driven by a 40hp steam engine, and could operate at up to two tons per sq.in. The supply pipes were of half inch bore with quarter inch walls.

Sections at the tube end showing the hydraulic presses for raising the Conwy tubular bridge

Fig. 362.

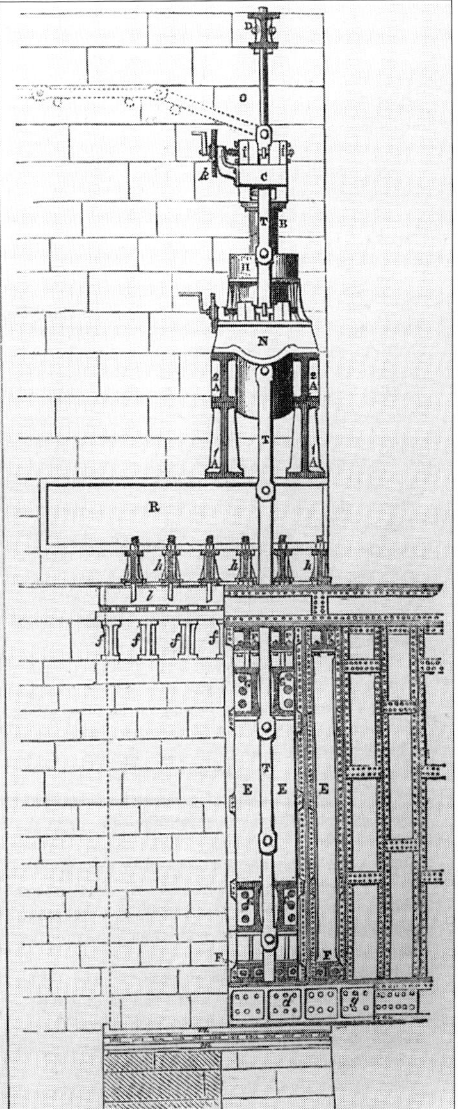

Fig. 363.

Onshore fabrication of the Britannia tubes

The design of the tubular bridges

In 1845, when Stephenson first conceived the idea of a wrought iron tube, acting as a beam, to form the deck of a bridge with a span of 450ft, such a large scale application of a material which had received little study was previously unknown. Only in the shipbuilding industry had anything remotely comparable in size been built, with the exception of two lattice girder railway bridges, constructed in Ireland by Sir John Macneill and Janes Thonson, which were of 150ft span. Although a brave attempt, experience with these bridges gave little confidence in their suitability for a span three times as great. Nonetheless, within a few years, this type of bridge was perfected, rendering the tubular bridge obsolete.

Fairbairn's extensive series of experiments proved what theory predicted - that a rectangular shape was most efficient in its use of material. The load carrying capacity of such an iron box section depended on (a) the capacity of the bottom to resist tension, (b) the capacity of the top to resist compression, these two forces being equal and opposite, and (c) the capacity of the sides to transmit any forces that might exist between the top and bottom. While the first two could be calculated accurately, the forces in the sides remained unknown in the current state of knowledge. The ideal tube would be so proportioned that each part was of sufficient strength to resist the forces it had to carry, neither more nor less. This was what Fairbairn set out to achieve. An important discovery made in his earlier experiments was that wrought iron is less able to withstand compression than tension. His experimental tubes, when made with a similar amount of iron in the top and bottom, invariably failed by a buckling of the top. However, earlier experiments on iron columns had shown that that a hollow tube was far stronger than a solid bar containing the same amount of material. By applying this principle to the top he was able to gain much increased resistance to buckling with little additional iron.

Having satisfied himself that he had acquired sufficient information to determine the overall dimensions of the full size tube, the next stage was to build a scale model one sixth the full size, which could be tested to destruction, and repaired and modified, until as nearly as possible the ideal section had been achieved. The bottom was formed of solid plates, and the top in the form of rectangular cells. Stiffeners were added to the sides as required to avoid buckling, but commercial pressures to get construction under way prevented any further refinement of the sides which might have been desirable. When a fair balance between the bottom and the top had been achieved, Fairbairn was satisfied.

The practical design of the full size tubes introduced certain considerations dictated by the knowledge of the day, and the limitations of the iron sections available. This particularly applied to plates, the size of which was limited by the weight which could be handled and the capacity of the rolling mills, and the inability of some ironmasters to supply the quality of iron required, particularly for the bottom plates. Both top and bottom were constructed in the form of rectangular cells, with plate webs and angle irons at each corner, and two layers of plates on the top and bottom of each cell. These thinner plates were proportionately larger and easier to punch, and made it unnecessary to rivet through more than four thicknesses, i.e. two plates plus angle and cover irons. Furthermore, since men could enter the calls, future maintenance was simplified.

The sides of the tubes were designed by engineering judgment rather than empirical results, and certainly not by theory for none existed until several years after completion of the bridge. Erring on the side of safety, about one fifth of the weight of the tubes lay in the side plates and another fifth in the tee iron stiffeners placed two feet apart along the tube.

Each tube was designed as a simple beam, and there is no evidence in Fairbairn's writings that he ever contemplated relying on the theoretical benefits of continuity to give the required strength. As he wrote, with reference to Pole's analysis - " *I have given the above investigation of Mr Pole rather to stimulate experimental inquiry than to imply reliance in his conclusions.... Now the formula for this calculation may or may not be correct as the premises on which it is founded approach or recede from the truth; it is apparently, however, not derived from direct experiment, but from assumed data, which may be questioned in our attempts to reduce it to practice.*"

Fairbairn's final test on the 1/6 scale model was the one on which he based his full size tube designs. The model spanned 75ft between supports, was 32in wide and 54in deep, and weighed 6 tons. It failed with a central load of 86 tons. From his earlier experiments Fairbairn had derived a formula linking the breaking weight to the cross sectional area of the bottom, the depth, the span, and a constant C which incorporated all the other factors and was specific to a given form of tube. For the model tube C was about 76 which he appears to have rounded up to 80 for convenience. In practice exact scaling up was impracticable, since in the interests of economy the top and bottom plates were made thickest towards the centre and the side plates towards the ends. In addition, tee iron stiffeners on the sides, intended to resist the possibility of buckling, added about 20% to the overall weight without any appreciable gain in loadbearing capacity. The Britannia tubes were also

Lattice girder railway bridge at Runcorn

increased in span from the original 450ft to 460ft clear span, and 472ft overall.

The computed weight of the Britannia tubes, by scaling up, would have been about 1400 tons. Clark quotes 1553 tons for each of the main tubes as built. The strength of the tubes was computed by Clark, giving a minimum central breaking load of 2537 tons, or 5074 tons evenly distributed. Deducting the tube weight of 1553 tons leaves 3521 tons. The customary maximum working load at that time was taken as 1 ton per ft run, and a factor of safety of six. The tube would in fact carry 7.65 tons per ft before breaking and appeared to possess ample strength.

However, as Professor Airy pointed out, the true factor of safety would be 5074 /(1553 + 460) = 2.52 instead of the 4 which became standard practice as a minimum in later years.

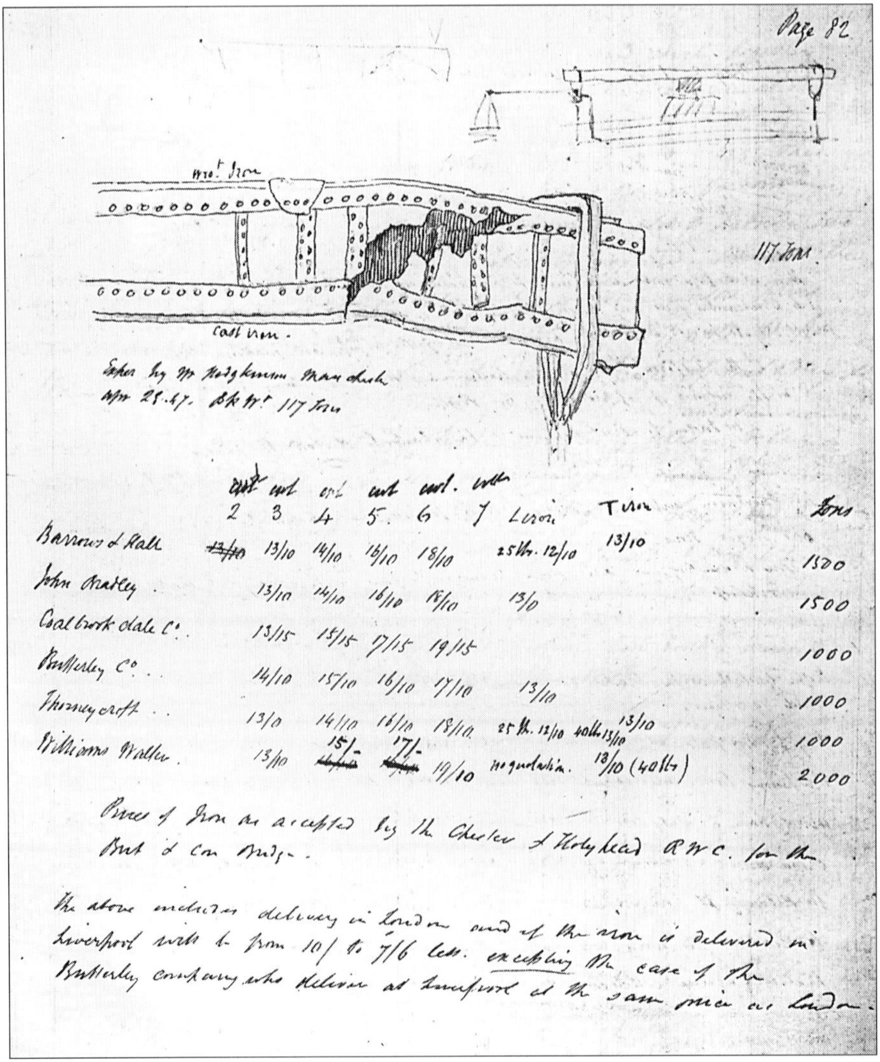
Page 82

Wro.t Iron

117 Tons

Cast Iron.

Broken by Mr Hodgkinson Manchester Apr 23.47. Bk Wt 117 Tons

	Cwt 2	Cwt 3	Cwt 4	Cwt 5	Cwt 6	7	Lever	Tires	Tons
Barrows & Hall	~~13/10~~	13/10	14/10	16/10	18/10		25lbs. 12/10	13/10	1500
John Bradley		13/10	14/10	16/10	18/10		13/0		1500
Coalbrookdale Co.		13/15	15/15	17/15	19/15				1000
Butterley Co.		14/10	15/10	16/10	17/10		13/10		1000
Thorneycroft		13/0	14/10	16/10	18/10		25lb. 12/10 40lb. 13/10	13/10	1000
Williams Waller		13/10	15/-	17/-	19/10		no gradation.	13/10 (40lbs)	2000

Prices of Iron as accepted by the Chester & Holyhead R W C. for the Brit & Con Bridge.

The above includes delivery in London and if the iron is delivered in Liverpool will be from 10/ to 7/6 less. excepting the case of the Butterley company who deliver at Liverpool at the same price as London.

Extract from Edwin Clark's diary

The maximum stress permissible in wrought iron at that period was very much a matter to be decided by an individual engineer. He would know that the maximum tensile stress was in the region of 18-20 tons per sq.in (tsi) although much higher values were bandied about. He would expect at least as high a stress in compression. It was many years later that the Board of Trade formally demanded maximum stresses of 5tsi in tension and 4tsi in compression. How did the Britannia tubes relate to these values? Clark gives 5.55tsi in the bottom with no load on the tube except its own weight. Loaded with 460 tons distributed, this figure increases to 7.19tsi, a very high figure indeed and, by later standards, indicating a woefully weak bridge. Of course, a load of 1 ton/ft would consist of locomotives nose to tail across the span, an unlikely event.

However, and this must be speculative, Clark probably realised that the position was not as bad as it appeared. By coupling the tubes rigidly end to end, there would be a degree of live load sharing between adjoining spans because there was perfect structural continuity in relation to any load in excess of the deadweight. It was a short step to realising that if the tubes were erected in such a way that they were bent in reverse over the piers the mid-span bending force could be much reduced, or heavier loads could be carried without strengthening. The next best thing to a free lunch!

There was one complication. Fairbairn's simple girders had not been designed to take a bending force at the ends - only at the centre, and it was far too late for any redesign with the first tube nearly completed. Fortunately there were heavy cast iron stiffening frames fitted at the ends of the tubes to resist the weight where they rested on the expansion rollers. In Clark's judgment, no doubt after seeking advice wherever he could find it, he could gain an additional 30% load carrying capacity without risking damage to the tubes.

How he set about this is described in some detail in his book, and is too long to repeat here. As Professor T.M. Charlton says: "*Certainly, they seem to have adopted a sound procedure on a scientific basis to which the long life of the bridge in increasingly severe conditions is adequate testimony.*"

Rivetting one of the Britannia Bridge tubes

Building Britannia

Work starts on the Conwy tubes

With the arrival of Spring in 1847 Evans made a start on the first Conwy tube, but only one ironmaster, Thorneycroft, was able to supply the large iron plates, 12ft long and 2ft 4in wide, needed for the bottom of the tubes. Such supplies as were available were sent to Evans, and Garforth and Mare at Menai had to delay their start until 8 August, when the first rivet was driven by Edwin Clark.

1847 was not to be a propitious year for the C&HR, nor indeed for many other well founded railway enterprises. During the previous year the Railway Mania had reached its peak, with 272 railway Bills before Parliament. The majority of these schemes would never see a yard of rail laid, but millions of pounds of shareholders' capital vanished into the open pockets of company promoters, speculators, surveyors and engineers of varying degrees of competence, and above all the legal profession. Never before was there such a lawyers' bonanza. In addition, the obsession of the Parliamentary Committees with competition meant that often two lines were encouraged and authorised where one could barely hope to pay its way.

Samuel Smiles said bluntly "*The result of the labours of Parliament was a tissue of legislative bungling ... Committees decided without judgment and without discrimination; it was a scramble for Bills, in which the most unscrupulous were the most successful.*"

The C&HR was more fortunate than most, in that, when completed, it would have an effective monopoly of the traffic and mails to and from Ireland, but that alluring prospect was several years away and meanwhile capital had to be raised. Many if not most shareholders were reluctant to answer calls on their shares. It was already clear that the final cost of the Britannia Bridge would far exceed Stephenson's original estimate of £300,000. Delay in starting on the Britannia tubes may not have been so much the fault of the ironmasters as the lack of money to pay them. Economies had to be made, and a start on the shorter land tubes was delayed until the following year, while the stagings on which the landward tubes were to be constructed were made narrower, allowing only one tube to be fabricated at a time. The staging was then dismantled and moved over to allow the second tube to be built.

Initially there were problems over riveting. Machine riveting had been called for wherever possible because of its superior results, but Mare, whose men were unhappy with the machines, resorted to using much heavier hammers giving greatly improved quality. At Conwy, Evans was apparently content with the steam riveters. Stephenson made a brief visit to Conwy on 22 July, and was generally satisfied with what he found, but instructions were given that all holes should be rymered (reamed). As he told Fairbairn "*This may not be necessary in every case; but as it is difficult to draw the line where the rymer should or should not be used, I thought it the safest plan to give an order that all should be done.*" The difficulty lay in aligning rivet holes through three or four thicknesses of iron plates and sections after the inevitable minor errors in punching.

Two months prior to this visit an event had occurred which threatened serious repercussions and could have destroyed Stephenson's hard won reputation. A train to Ruabon had been crossing the almost new Dee Bridge at Chester when a girder broke, and the train, except for the locomotive and tender, was precipitated into the river with the loss of five lives. At the Coroner's Inquest an eyewitness reported seeing the girder opening up beneath the locomotive, destroying the engineers' defence that the tender had been derailed and had struck the girder. The verdict of the Jury was that the bridge was too weak for the traffic, and Stephenson, as Engineer in Chief, was accountable.

A charge of manslaughter might and indeed should have followed, for local feelings were running high, but discreet steps must have been taken in the corridors of power to prevent this happening. A General Election was pending, and George Hudson (1800-1871), always anxious to have friends in high places, had effectively bought Stephenson an unopposed seat as MP for Whitby. Rather tactlessly rubbing salt into the wound, Fairbairn wrote to Stephenson on 10 June, extolling the virtues of the box girder bridge, and telling of one he was building for G.P. Bidder (1806-1878), Stephenson's friend and partner.

Evans had completed the first Conwy tube by the end of 1847, and it was decided that it should be tested for deflection under load before being floated out, in order to verify the results obtained from the scale model. The original scheme to stop up the ends and partly fill the tube with water was vetoed, since only sea water was available and serious corrosion problems would have followed. As neither Clark nor Fairbairn were able to make the computations, Eaton Hodgkinson was called upon to calculate the expected deflection. The test was carried out by placing the tube on temporary supports, and loading it with 236 tons of weights distributed over the central 300ft. This load gave a deflection some 20% higher than Hodgkinson had predicted. This was a matter of serious concern, since it was already known that the Britannia tubes with a span of 460ft were theoretically slightly weaker than those at Conwy and work on the first tube at Menai was well advanced.

Several of Fairbairn's learned friends had expressed a desire to witness the test but Stephenson, fearful perhaps lest something untoward should occur, insisted there should be no spectators. On completion of the test Fairbairn, for his part, considered the results to be satisfactory and incurred Stephenson's displeasure by communicating them to his friends by letter. Furthermore Stephenson chose to comment in a letter that Hodgkinson's experiments had confirmed the only true law, and that Fairbairn's results were merely empirical. The latter maintained a dignified silence, but was far from pleased.

The root of the problem lay in Fairbairn's empirical formula derived from the experiments on the model tube. Stresses in the top and bottom sections, and hence deflections, increased at a rate greater than the span. Neither Galileo nor Airy, let alone Fairbairn, seem to have appreciated the potential pitfalls in scaling up from models. There was a further problem in the method commonly used to estimate the strength of a bridge. The usual practice, based on the strength of cast iron, was to take the estimated central breaking load for a beam or girder bridge, deduct half the weight of the structure itself, and divide the result by a factor of six, or even more in the case of the most conservative engineers. The final figure was treated as the maximum safe working load.

In August 1846, acting on Stephenson's instruction, Clark had written to Airy with details of the proposed Conwy tubes, claiming on the above principle that the tubes would need to be loaded to seven times their maximum expected working load before breaking - a very good factor of safety in Clark's opinion. In his reply, Airy gently pointed out that this was not strictly true, and that the breaking load should be divided by the dead and live loads combined to arrive at the true factor of safety - in this case about 2.4, a very different matter. Stephenson was having none of this, and declared that he would adhere to his old system, and be prepared to work at a much lower factor of safety ! It was the following May that the Dee Bridge, with a true factor of safety of barely unity due to fundamental design errors, reduced even further by the addition of ballast under Stephenson's personal supervision, fell into the river.

The practice of the Board of Trade Railway Inspectors was to check the deflection of bridge girders under various loads, and the result of the Conwy test had suggested that there could be a problem at Menai. The bridges had been designed to carry a distributed live load of 1 ton per foot run, and in practice at that date the heaviest working train would be little over half of that. Consequently it was not primarily a matter of operational safety. There is no evidence as to what took place at this stage, but Clark's later admission that the Britannia tubes were weaker than those at Conwy suggests that there was considerable concern, although Fairbairn was adamant that his tubes were designed as simple beams and were of adequate strength for any load which might be applied.

At Conwy, should the unthinkable occur, there were the options of reverting to chains or, as was done many years later, of reducing the clear span by introducing extra piers. At Menai the second option did not exist, and the addition of chains would add enormously to the cost at a time of financial stringency. One possibility remained and may have been proposed by Stephenson himself as a result of his previous cantilever designs. The concept was simple enough, but it called for a mathematician of unusual ability to quantify it.

The benefits of structural continuity

It had long been known by carpenters and architects that a long beam resting on three or more supports would carry greater loads than short beams laid end to end. But no one had attempted to use this principle on such a large scale, and no engineer then living could assess the benefit. It had been decided at an early stage that the Britannia tubes would be joined into one length after erection and thus would form a physically continuous beam supported at five points. In this mode the dead weight of the tubes merely rested on the piers, and the full bending moment at centre span remained as if they were independent sections. If the tubes could be joined in such a way over the three piers that the top was permanently in tension, the tension on the bottom of the tube at mid-span would be substantially relieved.

At that time, early in 1848, there was possibly only one man in England who had studied the matter, and who had published a textbook for dealing with this and other structural problems. That man was Henry Moseley MA FRS (1891-1872), who had been appointed one of the first Inspectors of Schools. But to benefit from Moseley's calculations demanded a considerable knowledge of the calculus - a closed book to practising engineers at a time when a knowledge of higher mathematics was deemed unnecessary, even undesirable, as was suggested by C.B.Vignoles when appointed a Professor of Civil Engineering. Moseley was a personal friend of Fairbairn and so for Stephenson to seek help from him might have been embarrassing.

It is possible that Moseley's pupil, William Pole, a remarkable character in his own right, had acquired the necessary knowledge, but Pole had returned from India only at the end of 1847 and taken up employment with James Simpson CE, the eminent water engineer. A third possibility is that Clark's colleague on the C&HR, Charles Heard Wild, had acquired a sufficient understanding of the basic principles to make an informed estimate, just as he was to do a couple of years later with the Warren girder. Wild's "geometrical analysis", recorded in Clark's book, was not original, being lifted from the works of Duleau, and insufficient for the purpose. It is unlikely that the true facts will ever be known, but it is evident from Clark's own testimony that he failed to fully comprehend the business in which he was engaged and, when faced with complex calculations, he resorted to what was, at best, informed guesswork.

While this problem was under discussion, the first Conwy tube was floated out, raised into its final position, inspected and passed by Captain Simmons RE, and opened for passenger traffic on 1 May 1848. With Brunel in attendance, his friend, Captain Claxton RN had supervised the floating out and guided the tube into position ready for lifting. Clark, who had just lost part of a toe in an accident, was forced to watch the lifting of the tube from a carriage on Telford's suspension bridge. Stephenson drove the first locomotive to cross. But regardless of the rejoicings a major crisis was in the offing.

Fairbairn's resignation

Barely a fortnight after the bridge entered service, Fairbairn wrote to Stephenson suggesting that the time had come for him to resign. It can hardly have been a surprise for the relationship had become increasingly soured, and there had been rumblings from Fairbairn about resignation the previous May. What part Clark had played in fomenting the disagreement is open to speculation, but from his point of view Fairbairn's departure would mean that his own position would be greatly enhanced with no one but Stephenson to answer to. Clark had risen a long way in two years and now also enjoyed the benefit of having his brother as a personal assistant.

In fairness it should be said that it may have suited Fairbairn's book to resign at this stage, for his salary was a pittance compared with what he could earn as the master of an engineering works employing as many as 2,000 men. Moreover the Millwall shipyard had proved unprofitable and a constant drain on the Manchester enterprise. There was much that needed attention at home rather than pulling chestnuts out of the fire for an ungrateful Stephenson.

Fairbairn's letter, written at Manchester on 16 May, had probably not been received by Stephenson when he spoke at a celebratory dinner given at Conwy the following day, but Fairbairn took such offence at certain aspects of the speech, which was widely reported, that he immediately confirmed his resignation in a further letter on the 19th. When by the 22nd neither letter had been acknowledged Fairbairn wrote a formal letter of resignation to the Directors. To Fairbairn's discomfiture, on receipt of this letter the Directors reneged on their original terms on which they had made the appointment, and required the resignation to be submitted to Stephenson. It was all done in a very civilised manner, but there is little doubt of the pent-up emotions underlying the restrained language. Fairbairn was a proud man and felt his reputation had been publicly slighted. It was obvious from the tenor of Stephenson's speech that he, with a nod towards his "closet companion" Clark, was claiming for himself any credit going.

What particularly aroused Fairbairn's ire was a repetition of the specious claim regarding the bridge at Ware and priority for the original concept of cellular construction. This apocryphal structure was never built, and when pressed on the matter by Fairbairn, Stephenson produced a drawing of a primitive plate girder. Furthermore Fairbairn and Hodgkinson were given little credit for the vital part they had played in developing an ill formed concept into an accomplished reality. Clark on the other hand was rewarded with a fulsome tribute, but the audience was left in no doubt as to who deserved the lion's share of the credit - Stephenson himself.

Fairbairn's immediate reaction was to pick up his pen and write a book setting forth the story of the tubular bridges and his part in their creation. Publication caused a considerable furore both within the engineering profession and among a wider public. There were those who thought that Fairbairn deserved equal credit with Stephenson but there were others who felt that such a public disagreement was unseemly and did Fairbairn's cause no favours. A paper war developed between the soldier and journalist, Sir Francis Head, and Thomas Fairbairn, replying on his father's behalf.

Although the bridge was not without its critics, the public perception was that it was the great Stephenson's brainchild, and that he alone deserved the credit. Better informed observers thought that the honours should be shared, but the popular myth has prevailed over the years, and William Fairbairn has, as a result, been denied his proper place in history among the great engineers. In its review of Clark's book, the *Civil Engineer and Architect's Journal* took the view that Fairbairn was only a paid assistant and acting under Stephenson's orders.

The *Mechanic's Magazine* tells of an article in *The North British Review* attributed to Sir David Brewster, an eminent scientist and one of the earliest members of the ICE, giving undivided credit to Stephenson. The eminent author later confessed to being misled by a pamphlet published by Clark, and after studying the history of the bridges he unequivocally awarded the major credit to Fairbairn. The Editor commented:

"*To talk, after this. of Mr.Fairbairn's being entitled only to a secondary and subordinate place in the affair, is to outrage all truth and propriety. We can but regard with profound pity the hallucination which has betrayed a man of Mr Stephenson's genius and worth (this unfortunate episode notwithstanding) into so false a position.*"

In his Presidential Address to the Institution of Civil Engineers, Sir William Cubitt sought to heal the divisions in the profession. He expressed the opinion that a public conflict between two such senior members was damaging and that concessions should be made on both sides.

An article in the 8th edition of the *Encyclopaedia Britannica* published in 1856 was contributed by Stephenson, or at least bore his initials, although one of his assistants, G.D. Phipps, was believed to be the true author. Nowhere is there any mention of Fairbairn's contribution. Failure to offer any recognition eight years after Fairbairn's resignation shows that Stephenson carried this legacy of bitterness to the grave.

Writing after Stephenson's death in his *Industrial biography*, Samuel Smiles was more forthcoming than before, and stated:

"*There is no reason to doubt that by far the largest share of the merit of working out the practical details of those structures, and thus realising Robert Stephenson's magnificent idea of the tubular bridge, belongs to Mr Fairbairn.*"

G. Drysdale Dempsey, who wrote much on engineering matters, and who knew Fairbairn personally, was in no doubt, and wrote "*That these great works owe their design and construction to those joint labours is clearly evident, and, we respectfully submit, amply sufficient to justify the record of the two names of ROBERT STEPHENSON and WILLIAM FAIRBAIRN in an honourable and enduring association.*"

Edwin Clark takes charge

Fairbairn's abrupt resignation can have come as no great surprise to Stephenson. Possibly in some ways it was a relief for they were two proud men, both distinguished in their own fields of activity, and Fairbairn, fourteen years the senior, was unaccustomed to taking second place. Clark had proved a quick learner in his two years on the project, and Frank Forster and Alexander Ross, whatever they may have thought of Clark's meteoric rise, were very capable engineers and loyal to Stephenson. In William Evans at Conwy they had a competent and enterprising contractor, and Mare's senior foreman at Britannia, Thomas Fleet, clearly knew his business and work on the tubes was progressing satisfactorily. Fairbairn was no longer needed, as he himself realised, but a little more appreciation of his efforts would not have gone amiss in his opinion.

By the middle of 1848 work on the first two Britannia tubes was well advanced, but money was hard to come by. Although the Chester & Holyhead had emerged relatively unscathed from the financial crisis at the end of 1847, which stemmed in large part from the excesses of the Railway Mania, it had only been with the help of the London & North Western Railway. The difficulties in 1848 arose from heavy overspending, and Stephenson was called to account by the Directors who demanded economies. He was reluctantly obliged to order progress on the second line of tubes to be slowed, and instructions were given that scaffolding for only one line of the land tubes should be provided when the time came.

Mention should be made here of Captain Constantine Richard Moorsom RN (1792-1861). Although the Company, like their Engineer-in-chief, was based in London, the Directors wisely saw fit to appoint one of their number as Resident Director, based in North Wales, on a salary of £1,000 a year, and heading a local Works Committee. Moorsom took up his position in September 1844, and went to live at a house called Gorphwysfa overlooking the Straits. Moorsom's function has been described as that of a trouble shooter, and he acted as general manager of all aspects of the building of the line and the operation of the steamships. With Stephenson engaged elsewhere for so much of the time Moorsom undoubtedly more than earned his salary.

Perhaps his most onerous duty was the continual need to keep things on an even keel despite recurring financial crises, and his links as a Director also of the LNWR undoubtedly smoothed the relationship between the two Boards. To the LNWR the Chester & Holyhead

Floating out the one of the Britannia tubes

must have looked at times like a bottomless pit. In October 1848, when finances were once again at crisis point, Moorsom offered to forego half his salary. He was to receive his reward later, becoming in turn Chairman of the C&HR and the LNWR. Throughout all this he had remained on the Navy List on half-pay, and by 1857 had risen to the rank of vice-Admiral.

To add to Stephenson's problems, the death of his father on 12 August came as a severe blow. Although their relationship, particularly in earlier years, had often been strained, even

The Anglesey entrance showing two of John Thomas' stone lions

stormy, things had improved considerably since George had retired in 1840. Bidder no longer needed his diplomatic skills to smooth things over between father and son. George had taken great pleasure in his son's achievements, and his presence on occasional visits to the Westminster office would be much missed, but to the modern reader Robert's expressions of filial piety seem somewhat extravagant, even insincere.

Progress on the bridge was once again threatened in October 1848. The Company's bankers demanded repayment of an outstanding loan for £75,000. The principal cause of the Company's financial woes was the earlier overspending, which included a contribution of £200,000 towards the harbour at Holyhead, demanded by the Admiralty, and the acquisition of a small fleet of steamers to prise the Irish mail contract away from the Liverpool shipowners. In all another quarter of a million pounds was needed, and the prospects of raising this money through normal channels were remote. Once again recourse had to be made to the goodwill of the LNWR. Even then the financial situation remained dire, and the outcome was that not only were the contractors called upon to take part payment in debentures, but Stephenson himself had to accept shares in part settlement of his fees.

An early victim of the economy drive was the giant statue of Britannia destined to crown the Britannia tower. Fortunately perhaps she was never to be heard of again. But the four giant stone lions, guarding the portals of the tubes, escaped the financial purge. They were the work of John Thomas, sculptor and architectural draughtsman, who had been commissioned to carry out much decorative work in the new Houses of Parliament. "*These noble animals, which are of the antique, knocker nosed, pimple faced Egyptian instead of the real Numidian form, although sitting, are 12ft high, 25ft long, and weigh 30 tons*" wrote Sir Francis Head, obviously impressed more by their size than their beauty.

Undeterred by the shortage of money, William Evans, a man of some substance, soldiered on at Conwy, and the second tube was successfully floated out and in position on its piers by 8 December 1848, while Captain Simmons passed it for use on 2 January 1849. It was however to be another fourteen months before Stephenson could drive the first train over the Britannia Bridge.

The massive stone piers of Thompson's architectural masterpiece were rising fast under the supervision of Frank Forster, and it was hoped that they would be ready to receive the first tube by early summer in 1849. With an extra effort at the end of 1848 Clark was able to report completion of the side towers before the end of February the following year. The larger Britannia tower, less accessible on its rocky outcrop amidst the fierce tides of the Straits, took slightly longer, but it had reached well above the level of the tubes within another month. The stagings for the land spans had been erected, and work on the tubes was reported as well in hand in November 1848. Even this favourable progress failed to satisfy the demands of the Directors, who told Stephenson that they expected the bridge to be open by August. Faced with this unrealistic requirement Stephenson doubtless reflected that the Company needed him more than he did them at such a critical time.

Now that Garforth had finished his contract Mare had a clear run to complete the remaining two main tubes and the four short land tubes. For a colourful description of the construction stage it is necessary to record the words of Sir Francis Head. He had witnessed and had been impressed by the activities on site, but the riveting process by teams of five men particularly caught his eye. The construction of the top cells was made exceptionally difficult by reason of their small size, but this was 1848, and a ready solution lay to hand as he reports :

"*A tiny rivet-boy - we observed one little mite only ten years of age - in clothes professionally worn into holes at the knees and elbows - crawling heels foremost for a considerable distance into one of these flues as easily as a yellow ferret trots into a rabbit-hole, is slowly followed by his huge lord and master the holder-up, who exactly fits the flue, for the plain and excellent reason, that by Mr Stephenson the flue was purposely predestined to be exactly big enough to fit him; and as, buried alive in this receptacle, he can move but very slowly, he requires some time, advancing head foremost to reach the point at which he is to commence his work.*" Stephenson was clearly a thoughtful designer with his operatives' comfort close to his heart!

It was not until April 1849 that the C&HR Board gave their approval for work to proceed normally on the second pair of main tubes. Meanwhile within a month the first Anglesey tube was ready for floating out. It had taken some seventeen months to build, and another four months was needed to cut away the rock below the staging so that pontoons could be floated beneath the finished tube. The pontoons at Conwy, built by Evans the contractor, had been of somewhat lightweight timber construction, and Stephenson gave orders that heavier iron pontoons should be built for Menai, where the tubes to be floated were much heavier and the tides fiercer. The additional cost could be justified to Captain Moorsom and the ever vigilant Board of Directors since the iron pontoons would have a resale value.

The floating out was scheduled for 19 June, weather permitting, and there was a great gathering of spectators of all ranks of society and an atmosphere redolent of Derby Day on Epsom Downs. Sir Francis Head joined the throng of onlookers, most of whom, as he observed, were drinking, eating, watching various striking entertainments, and generally giving little thought to the serious matter ahead. Privileged spectators, including Vignoles and Bidder, were accommodated on top of one of the remaining tubes, whence Stephenson directed operations.

Brunel and Locke, who had come to offer moral support, volunteered their services on shore, and Captain Claxton was in charge of the sailors, 105 on each pontoon. The eight pontoons, capable of lifting 3,200 tons, nearly twice the weight of the load to be moved, lay ready beneath the first Anglesey tube at low tide. As the tide rose the tube lifted free of its supports, and had started to be hauled out, serenaded by a brass band, when one of the hauling-out capstans failed. Safely but ignominiously the tube was returned to its dock, and the crowds dispersed, hoping for better fortune the next day. They were not to be disappointed.

The following day all went smoothly, except that one of the land based capstans was wrenched from its foundations with potentially serious consequences. Fortunately the presence of mind of Charles Rolfe, the man in charge of the capstan, who was normally employed as an inspector of ironwork, saved the day by calling on the services of a crowd of spectators, and the hawser was secured by the efforts of the crowd. Within about two hours the tube was in its position between the two towers, and water was let into the pontoons in order to lower it on to its bed. The pontoons were eased out from under the tube, and duly returned to shore ready for the next lift. While the tube lay securely just above high water the lifting tackle was installed in the towers.

The two presses from Conwy were placed in the Britannia tower, and the new and more powerful press fixed in the Anglesey tower, but the large press leaked under pressure due to a porous casting, and nothing that Amos, the supplier, could do offered any cure. A new casting had to be ordered from the Vulcan Foundry at Warrington, which resulted in two months delay.

Finally on 10 August all was ready for lifting to commence. Stephenson could afford to take no chances of a failure in the lifting gear, and ordered that, as each lift progressed, the space below the end of the tube should be filled with brickwork and timber packing. The wisdom of this was soon demonstrated, for on 17 August the large press again failed, the base breaking away from the body, allowing the tube to fall nine inches. Clark and his brother were thrown down but had a miraculous escape. The only fatality was one unfortunate sailor. This time the casting was replaced in only six weeks, and the base was redesigned to avoid stress concentrations, which were the apparent cause of the failure. No further misadventures of any consequence occurred, and the tube reached its final resting place on 10 November. But while moving the press a week later a rope broke free from the capstan and the massive press fell into the water. Another sailor died, but the press, when salvaged, was found to have come to little harm. It was speedily transferred to the Caernarfon tower ready for the second tube to be lifted.

There had been worries at Conwy about water penetration into the top cells, and both tubes were roofed over with corrugated iron on an angle iron frame to Stephenson's satisfaction when he returned at the end of November to organise the floating out of the second Menai tube. With the experience gained previously there was greater confidence, and after one or two minor problems the tube was floated out on 4 December, and although one of the 12in hawsers had, in Clark's opinion, been sabotaged, the tube was safely delivered into place, greeted with a salute from a battery of field guns. Lifting proceeded uneventfully, and the second tube was home and dry by 7 January 1850. The land tubes had been completed on their stagings, and all that remained was to join the four tubes end to end.

There remained the question of obtaining structural continuity which Stephenson was told would increase the strength of the bridge by 30%. On the scant evidence available it would seem that the decision to follow this course, instead of merely joining the tubes end to end, may have been reached after the test on the Conwy tube had shown more than the expected deflection under load. To reduce the bending moment at the centre of the span it was necessary to lift the end of one tube relative to its partner before making the joint. Then when the end was lowered again it would set up an opposing bending moment over the pier. The short land tubes had been built initially with one end raised in preparation, but this meant that the amount of lift had to be known in advance. Clark is silent as to whom he turned to for this information. But there was one snag, for instead of making his first joint at the Britannia tower, which would have made the forces in the bridge symmetrical, Clark chose, for whatever reason, to join first the land and main tubes at the Anglesey tower.

He followed by making the joint at the Britannia tower in such a way that the reverse bending moment at the tower was intended to be about equal and opposite to that at mid-span. This, he admitted, gave less than perfect continuity, which could only have been achieved by a major redesign of the tubes at the central joint to accept the additional loading. In Clark's own words: "*The third and last junction made was that at B (i.e. at the Caernarfon tower): the calculation of this would be very complicated, but we are justified in assuming that the tube was made perfectly continuous at this point ...*" Certainly an effective degree of continuity was obtained, but it was far from symmetrical. Fairbairn conceded that continuity could be of some advantage, but insisted on the necessity for experimental work to quantify the benefits. His opinion was that mathematicians like Pole lived in an ideal world. In fact, Pole's impressive calculations, quoted in Clark's book, are the ideal which Clark failed to achieve in practice.

Regardless of theoretical objections, the first line of tubes was duly completed, the permanent way laid, and on 5 March 1850, with Stephenson driving the first of three engines, the bridge was safely crossed and recrossed. Emboldened by this, 21 wagons of coal were attached, the whole train weighing about 300 tons, and various tests were carried out successfully. On the stroke of noon Stephenson, expertly assisted by Mare, ceremonially drove the last rivet by candle light. Finally an enormous train weighing 503 tons and carrying 700 passengers, together with 45 wagons of coal, safely reached Anglesey and, minus the coal wagons, proceeded to Holyhead for the usual jollifications. That the Board of Trade had not yet passed the bridge for passenger traffic was of little consequence. Ten days later Captain Simmons unhesitatingly approved the bridge after testing it with a modest 228 tons. It had already been well tested!

Cheered by the opening of the line through to Holyhead and the prospect of increased traffic receipts, and despite their unresolved financial crisis the Directors took the plunge and ordered completion of the up line to proceed. The first tube was floated out on 10 June, and for all the experience gained earlier similar problems recurred. As before Stephenson and Brunel were in attendance, but this time the spectators were fewer and the operation went off quietly. The redoubtable Captain Claxton, with his speaking trumpet and flags, once again led his sailors to victory over wind and tide in only 50 minutes. Four weeks later the tube was in its final place.

The second tube followed on 25 July, and was in position on 16 August to the cheers of an immense crowd of onlookers.

It was not until 19 October that Captain Simmons was able to report on the completed second tube, and having satisfied himself he advised that it should be put into immediate use to obviate the dangers of the temporary single line working that had been in service for six months. This required the services of a pilot-man who rode on each train, hopefully returning with the next the other way. Failing this, he had to walk back through the sulphurous fumes in the tube, but small circles of glass had been provided in the sides to offer some degree of illumination for the unfortunate pedestrian.

The Directors complained of the heavy expense entailed and thought the precaution unnecessary, but reluctantly complied.

In his report, Captain Simmons accepted Stephenson's assurance that the use of continuity had reduced the maximum stress in the iron from 5.15 to 3tsi , whereas William Pole had calculated a maximum of 6.3tsi Stephenson's estimate was somewhat optimistic, but, as Simmons reported "*...having full reliance in the care and skill displayed in constructing this immense tube, I feel assured that every confidence may be placed in the security of the structure...*" The Dee Bridge failure three years earlier had established the principle that the Company and its engineer were ultimately responsible for the safety of their structures. Simmons contented himself with checking the deflection produced by a modest train of coal wagons weighing 228 tons rather than the 460 tons which each span was designed to carry.

Stephenson had been concerned from the outset regarding the effect any high winds might have on whatever bridge design was chosen. The deck of Telford's bridge had been severely damaged in a storm in January 1839. It was the reduced wind loading on a circular or elliptical tube which he had to offset against the superior strength of a rectangular one. To satisfy himself he set Clark to work to compute the wind load at a pressure of 20lb per sq.ft on the side of the tube. Clark concluded that the maximum stress that this would cause in the iron would be about half a ton per sq.in. Unusually, the natural period of vibration of the bridge when struck by a gust of wind was also assessed and found to be 67 cycles per minute, or 1.117Hz.

At this period there was much controversy over the effects of a moving load. Did it, as some argued, reduce the deflection the faster it went, or did it make any difference at all? There was an awkward minority who insisted that the vibration and impact of a moving load increased the stresses in the structure, and needed to be taken into consideration. In the hope of resolving the conflict of opinion, which in fact took another eighty years, Captain Simmons, who happened to belong to the awkward squad, carried out a series of trials on Sunday 20 May before the bridge opened. Unfortunately he depended on a water level to check the deflection of the tubes. Not the most precise of instruments even under static conditions, it was completely useless for the purpose in hand. Unfortunately the results apparently confirmed the views of the faster the better party, and the Captain was taken to task by the Press for expressing views contrary to natural law. Having, as the pundits observed, wasted the Sabbath, the trains were able to start running on Monday 21 October.

Model section showing the proposed corrugated iron roof

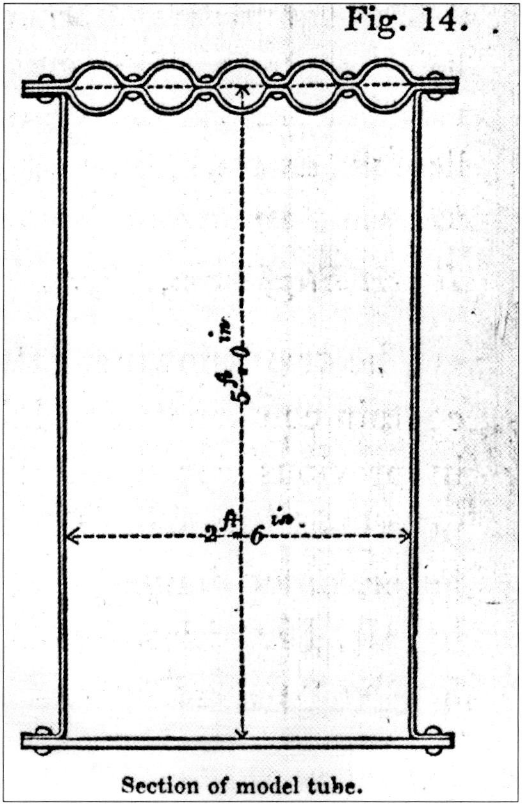

In practice the inertia of the tubes was so great that the small locomotives of the time would have had little effect, but a century later, with locomotives of four or five times the weight, impact loading was to become a potentially serious matter for concern.

Apart from the clearing up and the sale of surplus materials, there remained the matter of roofing the tubes to avoid corrosion. Stephenson had been criticised for his apparent extravagance in using corrugated iron at Conwy. In this case he was for some reason not consulted, and Clark, who had already left the Company for greener pastures, advised the

View of the completed Britannia Bridge from Anglesey

use of wood and paper as an economy measure. Later upgraded to tarred canvas on wooden boards, it was this highly combustible structure which would promote the rapid spread of the fire which engulfed the tubes in May 1970.

In the two and a half years following Fairbairn's abrupt resignation Stephenson had been under great pressure. Fairbairn's self confidence had complemented Stephenson's more cautious and less sanguine nature. On his own, and even with the support of Bidder and a trusted staff, Stephenson was to suffer his moments of black doubts, and the strain told on him. Beset by chronic health problems he increasingly turned for relief to alcohol and drugs, to the dismay of his friends and to the further detriment of his health, and at only forty seven years of age he was already past his prime. Quite apart from the Welsh bridges, he was deeply involved in two trunk lines, the Newcastle and Darlington and the Newcastle and Berwick, which formed the last two links in a chain of Companies between London and Edinburgh. The High Level Bridge at Newcastle and the Royal Border Bridge over the Tweed were major undertakings in their own right. The Resident Engineer, Thomas Elliott Harrison (1808-1888) bore the brunt of the work, with Stephenson himself, as he publicly confessed, merely exercising overall supervision.

At the end of August 1850 Stephenson was offered a knighthood by the Home Secretary, Sir George Grey, at the command of Queen Victoria, but he declined the honour, just as Fairbairn was to do eleven years later. They both felt that knighthoods were too often given to time servers or in return for political activities rather than to those who had rendered genuine public service. Some years before, at his wife's insistence, he had adopted a coat of arms, derived by some inventive herald from a distant Stephenson of Scottish origins, but he himself had never liked or ever used this pretentious emblem.

A request from the Directors that the Britannia Bridge should be formally opened by the Prince of Wales, then a boy of eight, was refused by the Prince Consort, but in October 1852 the Queen's carriage was drawn through the bridge by a team of workmen during a Royal visit to North Wales. Prince Albert and Edward, the young Prince of Wales, accompanied by Stephenson, walked over the top of the tubes, presumably on Clark's wood and paper roof.

The Britannia Bridge in service

Criticisms of the Britannia Bridge

No public project lacks critics, and the Britannia Bridge was no exception. The great tubes were a supreme example of form following function, owing nothing to any established style and mercifully free of the passion for applied decoration typical of the period. As a concession to aesthetics, as well as for economy, the tubes at Menai were tapered from the Britannia tower to the abutments, the soffit being level but the top on what was claimed to be a parabolic curve. This was compatible with good engineering practice and not merely cosmetic.

The architect to the C&HR was Francis Thompson, who had cooperated successfully with Stephenson previously. If not to be classed as a great architect, Thompson was certainly a highly competent one with special experience of railway work, and his station at Chester ranks among the great stations of Britain. He had the ability to adapt any architectural style to the needs of the situation, and his choice of a simplified Egyptian Revival style for the towers at Menai complemented the stark simplicity of the tubes. The Egyptian Courts at the Crystal Palace in 1851. by Owen Jones and Joseph Bonomi, were a particularly fine example of the style. At Menai the massive towers resembling Egyptian temple pylons were admirably suited to the great weight of the tubes, and the suspension chains which were originally regarded as essential auxiliary supports. Needless to say, the towers were not universally admired in an age when simplicity was too often regarded as a defect, but the substantial appearance of the bridge offered reassurance to a public suspicious of large iron structures.

At Conwy the engineers' efforts were less successful due to the close proximity of the medieval castle, and Telford's graceful suspension bridge. While Thompson designed the end towers to harmonise with the castle, not entirely successfully, the tube itself could not be disguised. Moreover, the increased height at mid-span, provided for sound structural reasons, might have been less obtrusive if hidden by a curve. Even painting the tubes a tasteful shade of light stone failed to soften the views of a correspondent in *The Builder* who proclaimed "*Its unmitigated ugliness is much to be deplored*". Fortunately perhaps, proposed decorative machiolations and plaques on the sides of the tubes failed to materialise. Probably, like the giant statue of Britannia at Menai, they were victims of a pressing need to economise.

From their first public announcement the bridges were beset by prognostications of doom. Not only from the public in general but also from those who regarded themselves as the highest scientific authorities, to the extent that Stephenson's faith in the project was at times sorely tried. From the Commons Committee to the oft-quoted "man on the Clapham omnibus" the tubular bridge was regarded by many as a dangerous novelty. With a sideways glance at Brunel's extravagant habits with shareholders' money, the gentlemen of the City considered that Stephenson should have adhered to tried and trusted methods, overlooking the obstacles placed in his way by the Admiralty. In fact, in reverting to the beam, Stephenson had chosen one of the most basic structural elements. The principle was unexceptionable but the scale of its proposed application was staggering. Only Fairbairn seems to have tackled the problem with unbounded optimism. Certainly Stephenson had his dark moments of doubt verging on despair.

The chief criticism of the tubular bridge, with hindsight a perfectly valid one, was that it was wasteful in the use of material, and that the extra weight of iron reduced the useful load that it could carry. Ten years later, nobody but Stephenson would have contemplated a tubular bridge on the scale of the Victoria Bridge, Montreal, such had been the pace of development. But in 1845-50 no one understood for sure the forces in the sides or web of a girder, and still less could they evaluate them. Perhaps Fairbairn should have extended his experiments on the model tube by reducing the thickness of the sides until they failed, and then assessing the benefits of web stiffeners, but he provided stiffeners anyway as part of his design philosophy. The pressure was on to build a bridge as soon as possible and it stood firm for 120 years and carried the traffic of the 1970s without protest. If 21% of the weight was in the side stiffeners and another 20% in the side plates, the extravagance paid off handsomely.

There were foreign critics, wise with new knowledge and the benefits of hindsight, who were able to point out faults. The great Clapeyron, writing in 1857, acknowledged the Britannia Bridge as being "of originality and grandeur", but goes on to say "This magnificent structure could be improved by some changes in the thickness of the iron plates, which seem to be relatively too weak at the supports". He failed to recognise that the tubes were never designed with continuity in mind, and contained no provision for reverse bending over the supports. Clark had to do his best in the circumstances. Molinos and Pronnier, who had taken the trouble to study the bridge as built, established that Clark had not succeeded in achieving the balanced distribution of stresses for which he had aimed. William Pole's mathematical analysis in Clark's book was a smoke screen, and written after the first tube was completed.

Away in Russia D.J. Jourawski, a practising railway engineer, developed a method of analysing trusses of the Howe type, and established a method of calculating the shear stresses in the webs of beams. This latter investigation was published in Russian, but not in Western Europe until it appeared in France in 1856. Jourawski pointed out that the most effective position for the stiffeners of the tubular bridges would have been at 45 degress to the vertical when half the amount of iron would have sufficed. Nine years too late to be helpful.

John Fowler's Torksey tubular box girder bridge

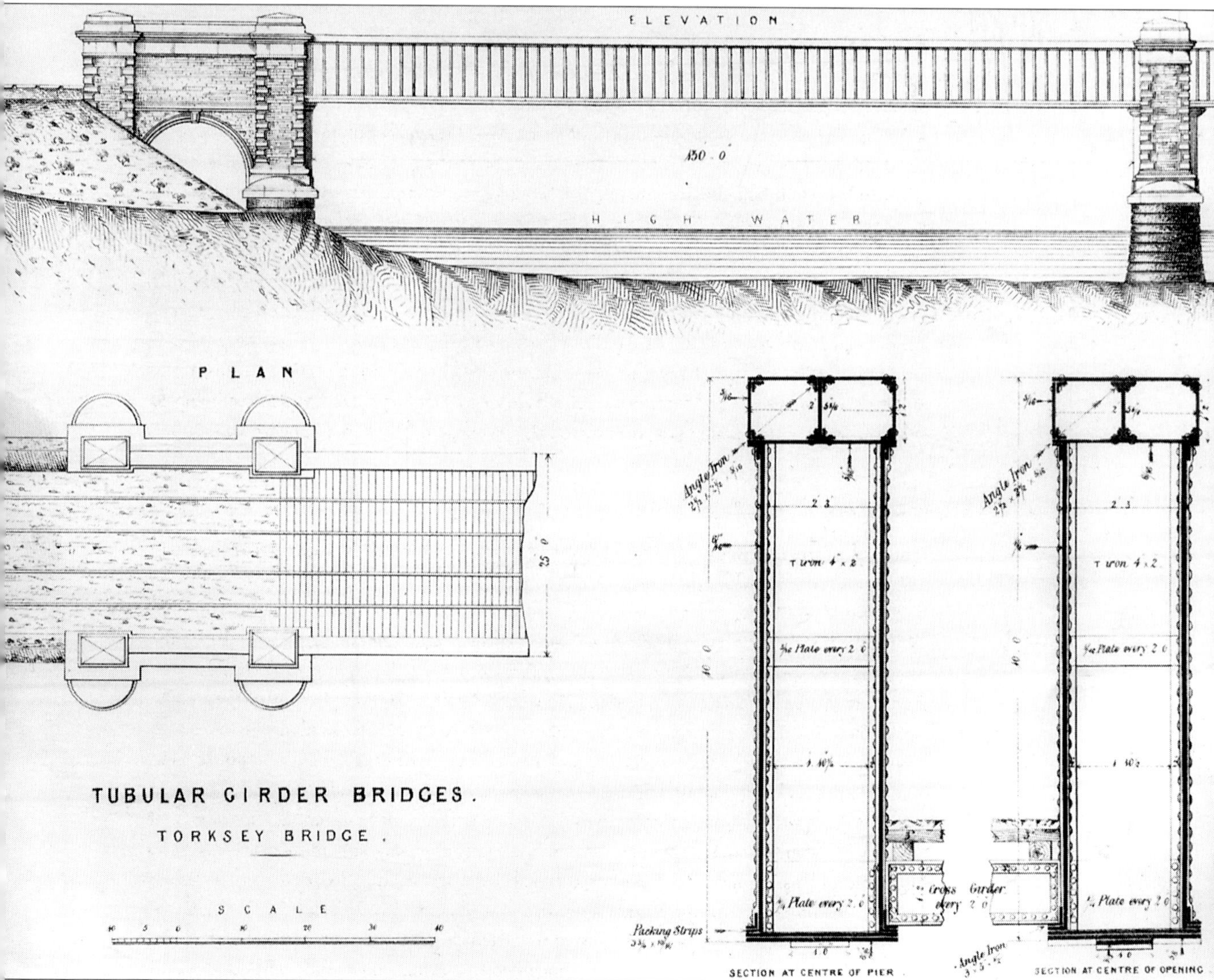

Other tubular bridges

In 1845 wrought iron bridges of any size were almost unheard of, but within five years the Conwy and Britannia bridges had established wrought iron as the supreme material for long span railway bridges. The opening of James Barton's Boyne Viaduct in Ireland in 1855 saw the open web girder triumphant over the immensely heavy solid iron plate sides of the train sized tubular bridges. In 1845, faced with a desperate situation and no precedent to guide him, Stephenson had fortuitously conceived the tubular form of beam, albeit with suspension chains. His doubts as to whether or not such a beam might be self supporting were finally laid to rest at Fairbairn's insistence. Had structural knowledge been further advanced the tubular concept might have been refined, but when that became possible about 1856 the development of open web girders had already rendered the tubular concept obsolescent. Consequently, in spite of his advocacy, Stephenson only built five more tubular bridges, and in the case of his three Egyptian bridges the rails were carried on top of the tubes rather than through them. However, it is unlikely that Stephenson was much involved with the latter, as the design is attributed to his cousin, George Robert Stephenson.

The Brotherton tubular bridge

The first of the five was at Brotherton over the Aire, and the design is attributed to Edwin Clark, although he was working under Stephenson's nominal supervision, and with the experienced J.C. Birkinshaw as Resident Engineer. The bridge formed part of the York and North Midland route between Burton Salmon and Knottingley, and was used also by the trains of the Great Northern Railway since it formed part of the main line between London and York. There were, as before, two single line tubes each 237ft long. In view of the smaller span Clark dispensed with the top cells and for some inexplicable reason made the tubes with only 9ft 3in internal clearance at the top and a foot more at the bottom. This did not seem to trouble the Inspector, Captain Wynne, who passed the first tube for traffic on 2 July 1851.

The GNR was far from happy with the situation, and complained that their latest type of brake van, when fitted with side lamps, had a clearance of only half an inch on each side. Apprised of this, Captain Wynne refused to allow the second tube to be opened, although his powers did not allow him to do anything about the one already in service. When asked for his comments, Stephenson merely listed restricted clearances on other railways which, he claimed, had caused no problems. These included one bridge, on the Manchester & Leeds

Railway, built by his father, where three passengers had been killed. Not for nothing was the instruction given "Do not lean out of the window". The dispute rumbled on, and Stephenson laid the blame on the GNR for using unreasonably wide rolling stock. In the middle of 1852 the two Companies compromised, and agreed to the widening of the tubes, sharing the costs with Stephenson, who undertook to supervise the operation. The top of each tube was split down the centre and the sides forced apart, and a plate 21in wide was inserted in the gap. This somewhat makeshift solution lasted fifty years, but cost Stephenson £642 as his share. The cost to his reputation was probably far greater.

The Victoria Bridge, Montreal

The last and the longest of the great tubular bridges was out of date even before work started. The Grand Trunk Railway of Canada had been projected to connect the Atlantic seaboard with the Great Lakes. The St Lawrence at Montreal is nearly two miles wide, but of no great depth and with a bottom of boulder clay on rock. Unfortunately to offset these advantages there was the enormous pressure exerted by massive ice floes, formed as the ice broke up in early Spring and was driven downstream by the force of the current fed by the melting snow.

After working on the plans for the Chester and Holyhead and supervising the construction of the Conwy Bridge, Alexander McKenzie Ross had gone out to Canada at the instance of certain speculators to spy out the land, and there he met John Young, a Canadian engineer, who had already proposed bridging the St Lawrence. Having sold the idea of a tubular bridge to Young, Ross returned to England to consult Stephenson, who visited Montreal in August 1852 for the first and only time, apart from a walking tour made on his way back from Columbia. Pronouncing the scheme practicable, Stephenson was appointed by the Company as Consulting Engineer for the bridge, with Ross as Resident Engineer, and Jackson, Brassey, Peto & Betts, the projectors of the line, as contractors. Ross himself was appointed as engineer in chief for the rest of the GTR.

After the engineer George Harrison (1815-1875) had visited Canada to study the situation at first hand, and on his advice, the contractors set up the Canada Works at Birkenhead to prefabricate ironwork for the bridge, and to manufacture other railway material for export, initially to Canada but later elsewhere. Harrison was appointed as Manager. The plans were

The Victoria tubular bridge over the St Lawrence at Montreal

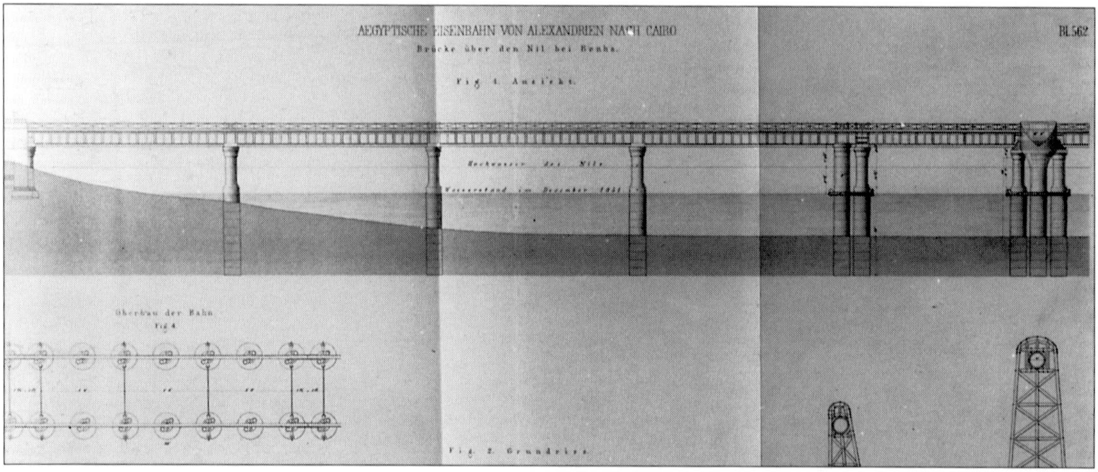

Robert Stephenson's box girder bridge over the Nile at Benha

prepared under the supervision of Stephenson's cousin, George Robert Stephenson with the detailing of the ironwork and the working drawings being carried out by B.P. Stockman. Later, Ross, who had done the initial site survey, was to claim credit for the design of the bridge, but it is probable that he was responsible only for the piers, no small undertaking in itself. Francis Thompson, architect of the Britannia Bridge, had emigrated to Canada, and was closely involved with the stonework of the Victoria piers, and all the stations on the Grand Trunk Railway.

The first stone was laid at Montreal on 22 April 1854, but such were the problems created by the cold in winter and fierce currents in summer that it was not until five and a half years later, on 24 November 1859, that the first train was able to cross the Victoria Bridge, nearly two months after the death of Robert Stephenson. The formal opening by the Prince of Wales took place in August 1860.

The bridge had twenty five spans, 24 of 242ft and a navigation span of 330ft allowing a 60ft headroom. The tubes were erected in position on scaffolding from ready machined and punched components sent out from Birkenhead. For greater accuracy holes were punched undersize in the factory and reamed out on site, but the tubes were dogged by faulty riveting throughout their life, exacerbated by the extreme cold of winter which made the metal brittle. The overall length of the tube was 6,588ft with the remainder of the crossing formed from stone faced embankments. From the start, despite the generous provision of vents, the bridge was plagued with smoke problems and unpopular with operating staff, but its demise was finally brought about by severe corrosion due to the constant spillage of brine from refrigerated freight cars, and the need for doubling of the track. The piers remain in service and have now resisted the ice for 140 years, although pressure grouting had to be resorted to in 1942 when signs of ice damage became apparent.

The Victoria Bridge attracted much criticism, mainly on two counts. The cost was £1,500,000 or not far off one hundred million pounds at current prices. The interest alone was to prove a millstone around the neck of the Company for years ahead, for the Grand Trunk Railway in its early years failed to attract the volume of traffic its promoters had anticipated.

An equally telling criticism was that open web girders had already become established by the time the Victoria Bridge was planned, and offered a far cheaper solution. Stephenson was never reconciled to the trussed girder, something he had never used and of which he had no experience, and in this matter the quality of his advice to the GTR was questionable.

There was an extensive use of tubular girders elsewhere on the GTR system. Detailed information is lacking, but the probability is that they were similar to the Egyptian bridges, with a rectangular box girder carrying a single line of rails on the top chord. Only in one case was the use of the through type of box girder resorted to for three spans. In 1857 Sir Charles Hutton Gregory (1817-1898) visited the GTR for a tour of inspection, and was favourably impressed with the quality of the tubes and girders. The Credit Viaduct, west of Toronto, with eight spans of 96ft caused him to comment "*I have rarely seen a work of finer design and execution.*"

Stephenson's Egyptian bridges

The only other tubular bridges built on Stephenson's orders were three on the Alexandria and Cairo Railway, and these were not of the through type but plain box girders with the track carried on top. The route had been surveyed in 1850/51 by Henry Swinburne who later became Section Engineer on the southern section under M.A. Borthwick, the Resident Engineer. A contract was signed between Abbas Pacha and Stephenson on 12 July 1851, and work on site commenced the following year.

In traversing the Nile Delta three major navigable waterways had to be crossed, the Rosetta and Damietta branches of the Nile and the Karrineen Canal. Initially, bridging the Rosetta branch was deemed impracticable for lack of firm foundation conditions, and steam ferries were planned, but work went ahead on the Damietta branch near Benha and the Karrineen Canal at Birket el Saba. At Benha there were eight box girder spans of 80ft and 6ft6in deep, with a central swing span of 154ft, providing two clear channels of 60ft each when open. Wrought iron tubular piers of 7ft diameter were sunk 35ft below the lowest water level. The bridge over the canal at Birket el Saba was similar in construction but smaller, with two spans of 70ft and two channels of 43ft each when the swing span was open. The design was attributed to George Robert Stephenson, and his assistant, George Barling. The contractor for the line as a whole was E. Price of London, and all the ironwork for the bridges was supplied by Grissell & Co., also of London. Both bridges were started in May 1853 and completed by October 1855. Some 10,000 pressed labourers were employed on the railway, working under an armed military guard to prevent desertion.

Possibly in the light of experience gained at Benha and the inconvenience of the ferries, it was decided in 1857 to construct a similar but larger bridge at Kaffre Azzayat on the Rosetta branch, the railway by that time having reached Suez. On this occasion the piers were of 10ft diameter, and sunk 60ft below the bed of the river using Hughes' compressed air process. The design was again the responsibility of G.R. Stephenson, assisted by B.P. Stockman.

There were nine spans of 114ft and the swing span, over 200ft in length, provided two 80ft channels. The bridge took only seventeen months to build, and was completed a few months before Stephenson's death. The Resident Engineer was a Mr McLaren, possibly the same John McLaren who supervised the ironwork at Britannia and Conwy. The contractor was again Edward Price of London, and all ironwork was shipped out from the Canada Works. The bridge was opened on 25 May 1859.

Fairbairn's tubular (box) girders

William Fairbairn was quick to realise the potential of the box girder for shorter spans, and, with Stephenson's somewhat reluctant agreement took out a Patent in 1846. During the next sixteen years over a hundred box girder bridges were constructed, and others

Fig. 10.

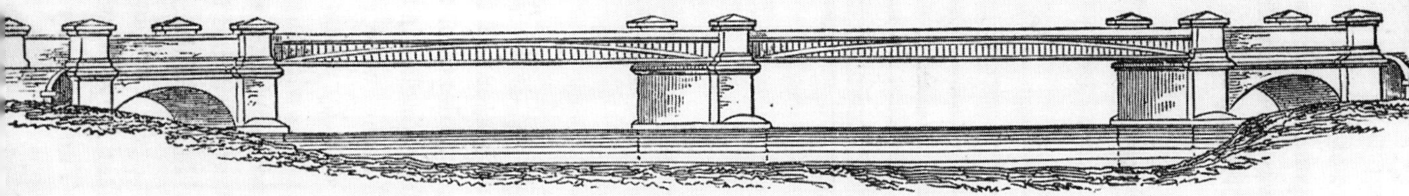

John Fowler's tubular girder bridge over the Trent at Gainsborough

continued to be built abroad up until about 1868. The first two were supplied to the order of C.B. Vignoles, never one to shun innovation, and were of a somewhat primitive construction. Three girders carried a double line of railway, and underslung timber cross beams supported the permanent way. Similar structures followed, one on the East Lancashire Railway, and two on the Liverpool & Bury, where James Thomson was engineer, and built the first wrought iron lattice girders in England.

A more sophisticated design was supplied to John Fowler who used box girders extensively on the Manchester, Sheffield & Lincolnshire Railway east of Sheffield. One of these, over the Witham on the southern outskirts of Lincoln, was opened at the end of 1848, and remains in use today after strengthening. The largest was over the Trent at Gainsborough, which had two spans of 150ft and opened in May 1849. As no obstruction of the waterway was permissible, this bridge was assembled on the bank and rolled out in one piece. A similar procedure was used a few miles upstream for the Torksey Bridge, which has two spans of 130ft. When inspecting this in December 1848, Captain Simmons took exception to, among other things, the maximum stress of 5tsi for wrought iron being exceeded, or so he claimed.

Simmons was correct if the two spans were to be regarded as independent beams, but he was on uncertain ground as the figure of 5tsi had no statutory significance. The ICE devoted four evenings to discussing the matter, after Fairbairn had read a paper on "Tubular Bridges". This was not altogether well received, and some of the young lions of the profession made their views known. Stephenson did not attend. With the aid of Wild and Pole, Fowler was able to prove that the bridge, as a continuous beam, was of adequate strength. Simmons did not yield easily, despite his unquestioning acceptance of the principle at Menai, but the

Railway Commissioners were pressurised into allowing the bridge to open, on the principle that it was the Company and their engineer who were ultimately accountable for any mishaps. The bridge still stands, unused and neglected, but well worthy of preservation as an engineering monument to Fairbairn and Fowler.

The last two major bridges to be built by Fairbairn were at Etherow and Dinting Vale, again on the MS&LR, where they replaced two large timber arch viaducts built by Joseph Locke, with Alfred Jannistreet Jee as Resident Engineer. Two years later Fairbairn's patent expired, but John Fowler continued using a similar design with cellular tops in Australia until 1868, shipping out the material from the Canada Works. Several were built in Ireland over the years, at least one of which is still in service.

New developments in iron bridges

A new generation of formally trained engineers was coming to the fore in the 1850s, several having been educated at Trinity College, Dublin, which in 1841 was the first University in the United Kingdom to have a School of Engineering. It was in Ireland too that the first iron lattice girders were built, and it was at Queen's College, Galway that the forces in the open web Warren girder were first evaluated experimentally. It was also in Ireland that Richard B. Osborne built the first iron Howe truss bridge in the UK. These young engineers regarded the tubular bridge as an engineering dinosaur, and its protagonists likewise. There was talk of the figures being cooked in favour of the tubular concept. A particularly vocal critic was Zerah Colburn, editor of *The Engineer* and founder of *Engineering* in 1866. Not only did he inveigh against the weight and cost of tubular girders but went so far as to favour a suspension bridge in the light of the success of his fellow countryman, John A. Roebling, in building such a bridge at the Niagara Falls in 1855 for the Great Western Railway of Canada. Colburn argued that a 3mph speed limit, as at Niagara, was a small inconvenience if the cost of the bridge could be halved.

In the 1850s a wide variety of single web girders were developed by Brunel and others, having the top chord in a tubular form, but the weakness of many of these, Brunel's included, was the failure to recognise the need to provide adequate stiffening for the web. By 1855 Fairbairn was producing single web plate girders closely resembling the modern configuration. At the same date the success of James Barton's open web lattice girder bridge across the Boyne, with a clear span of over 250ft, set the pattern for the future. Brunel's undeniably magnificent bridge across the Tamar at Saltash, with its two spans of 455ft each overall, was a brave and successful attempt to develop a long span bridge to rival the Britannia, but the complexity of its design offset the great reduction in weight, and the final cost showed only a limited saving. The future lay with the triangulated truss, modified later into the cantilever form for long spans.

The single triangular or Warren truss, patented in 1848, was never popular in the United Kingdom, but in its modified form, the double triangular truss, became widely used after its first introduction by R.H. Bow and Thomas Bouch in the famous Belah and Deepdale viaducts. This form of bracing is conspicuous in the Forth Rail Bridge. American truss designs were slow to be accepted in this country. The first Pratt truss was exported from London to Queensland in 1864, yet it was to be another twenty five years before the first Pratt appeared in the UK on the internal viaduct of the Forth Rail Bridge. In 1858 J.H. Latham, assistant to Janes Meadows Rendel, designed a bridge for India identical in principle to the American Whipple-Murphy trusses chosen, for the first time in the UK, by W.H. Barlow for the second Tay Bridge twenty nine years later. It may reasonably be claimed that while around 1845 to 1855 the UK led the world in iron bridge development, for the next thirty years there was a period of stagnation at home where railway bridge building was dominated by Company engineers. British consultants working abroad were far more receptive to new ideas, and open to American influences.

Artist's impression of the replacement Britannia Bridge

Epilogue: The Phoenix arises

1970 was to prove a fateful year for the Britannia Bridge. It had stood for 120 years, receiving the care which all iron bridges demand, particularly in close proximity to salt water. Apart from a regular programme of painting, inside and out, and the replacement of faulty rivets it imposed no great demands on its successive owners. The canvas and wood roof over the tubes received liberal applications of tar, and the inside of the tubes too, but after more than a century of steam engines passing through the possibility of a serious fire must have been regarded as remote. But old birds' nests and other debris had been allowed to accumulate within the roof over the years, and the bridge had become the proverbial tinderbox.

Of far greater concern to its owners as the years passed had been the ever increasing weight of locomotives and rolling stock, and the demands of the traffic department for heavier trains. When opened in 1850 the bridge had been considered more than adequate to support the heaviest train that could be put upon it, in the language of the day. In practical terms this was around 300 tons, although in theory it should have carried a distributed load of 460 tons on each main span - 1 ton per foot run. But in those days a heavy locomotive weighed at most 30 tons whereas by 1970 a locomotive weighing over 120 tons was not uncommon.

With the introduction of 102 ton gross wagons in the 1960s, the strength of the bridge was again called into question, since the original conditions might have changed during a century, and earlier assessments had been based on calculations alone. A series of actual weighings of the tubes at the points of support were carried out by British Rail around 1969. Wind loadings were also checked as these, originally of little consequence in Stephenson's day, had become critical with the increasing load. As a result of these tests it was concluded that the 102 ton wagons could be permitted to cross the bridge, but on 23 May 1970 the tubes were damaged beyond repair by a fierce fire, fanned by a strong wind, and within a few hours, with the tops of the tubes glowing red hot, the problem became of purely academic interest.

The cause of the fire was attributed to trespassers seeking birds' nests in the roof structure rather than arson, although the effect on the bridge was the same. The result of the fire was to cause the tubes to break apart at the side joints over the piers as they cooled. Although the fire had not affected the interiors, the main tubes had sagged over two feet at midspan whereas when new they were level at the soffit.

Reconstruction of the Britannia Bridge in the 1970s

It was decided to replace the tubes entirely with new steel arches spanning between the existing piers. This would not only cater for modern loading conditions, but would also allow a road deck to be built over the rail tracks. Fortuitously therefore British Rail were relieved of the responsibility of a slightly suspect and elderly if historic bridge, and the road authorities were able to provide relief for Telford's suspension bridge at a fraction of the cost of a stand alone structure. It was indeed an ill wind...!

Aesthetically the replacement is just another large steel arch bridge of no particular visual merit, although technically a fine example of modern structural engineering. The retention of the towers, disembowelled to accommodate the new road deck, does little to retain the spirit of Francis Thompson's grand conception, and it might have been a lesser act of vandalism to have reduced them to deck level. They stand as a sad reminder of what was once one of our greatest engineering landmarks.

THE

BRITANNIA AND CONWAY

TUBULAR BRIDGES.

WITH

GENERAL INQUIRIES ON BEAMS

AND ON THE

PROPERTIES OF MATERIALS USED IN CONSTRUCTION.

BY

EDWIN CLARK,

RESIDENT ENGINEER.

PUBLISHED WITH THE SANCTION, AND UNDER THE SUPERVISION, OF

ROBERT STEPHENSON.

IN TWO VOLUMES.

(WITH PLATES IN FOLIO.)

VOL. I.

LONDON:

PUBLISHED FOR THE AUTHOR,

BY DAY AND SON, GATE STREET, LINCOLN'S INN FIELDS;

AND JOHN WEALE, 59 HIGH HOLBORN.

1850.

Edwin Clark, The Britannia and Conway Tubular Bridges. *Cat. No. 20*

Afterword

On behalf of the Archives Panel of the Institution of Civil Engineers I would like to thank John Rapley for all the time and effort he has put into this essay. The Britannia Bridge was one of the great examples of British civil engineering of the nineteenth century, whose lessons, particularly with regard to buckling, had a great relevance to engineers designing steel box girders more than a century later.

The Britannia Bridge was a team triumph and perhaps I could also thank Rob Norridge for his help in designing this book and the Science Museum and Julia Elton for loan of material.

Mike Chrimes, Head Librarian, Institution of Civil Engineers, 1999

List of exhibits

1 W.A. Provis. An Historical and descriptive account of the suspension bridge constructed over the Menai Bridge, 1828

2 H. Parnell. Treatise on roads, 1838

3 T. Telford. Reports to Commissioners of Holyhead Road

4 Memorial of the promoters of the Chester and Holyhead Railway, 1840

5 Reports of the Committee appointed by the Lords of the Treasury relative to the best means of communication between London, Dublin, Edinburgh and Glasgow, 1840

6 J. Walker. Report and plans on communications with Ireland, 1843

7 J.M. Rendel. Report [on the] Chester and Holyhead Railway, 1844

8 Mackenzie and Brassey. Contractors ledger for the Chester and Holyhead Railway

9 W. Mackenzie. Diaries 1844-1850 relating to Chester and Holyhead Railway

10 R. Stephenson. Chester and Holyhead Railway - Reports ... upon the experiments made to ascertain the practicability of erecting a tubular bridge across the Menai Straits, 1846

11 [Horseley Ironworks]. Details of girders for a bridge over the River Dee, 1847

12 Illustrated London News. Dee Bridge failure, 1847

13 Commissioners of Railways. Copy of a report ... on the fatal accident on the 24th May 1847, by the falling of the bridge over the River Dee, 1847

14 Parry's Railway Companion from Chester to Holyhead, 1848

15 [S. Russell]. Penmaenmawr Viaduct, tunnel and sea wall, lithograph, 1849

16 General description of the Britannia and Conway tubular bridges on the Chester and Holyhead Railways. By a resident assistant, 1849

17 Civil engineer and architects journal, 1850. Report on testing of Britannia Bridge

18 Edwin Clark. Britannia and Conway tubular bridges: diary no.2, 1846-1850

19 Edwin Clark. Autograph correspondence with Robert Stephenson and Eaton Hodgkinson, 1846-1847

20 Edwin Clark. Britannia and Conway tubular bridges, 2 vols and atlas, 1850

21 G.D. Dempsey. Rudimentary treatise on tubular and other iron girder bridges particularly describing the Britannia and Conway tubular bridges, 4th ed., 1877

22 Thomas Fairbairn. Britannia and Conway tubular bridges: truths and tubes on self-supporting principles, 1849

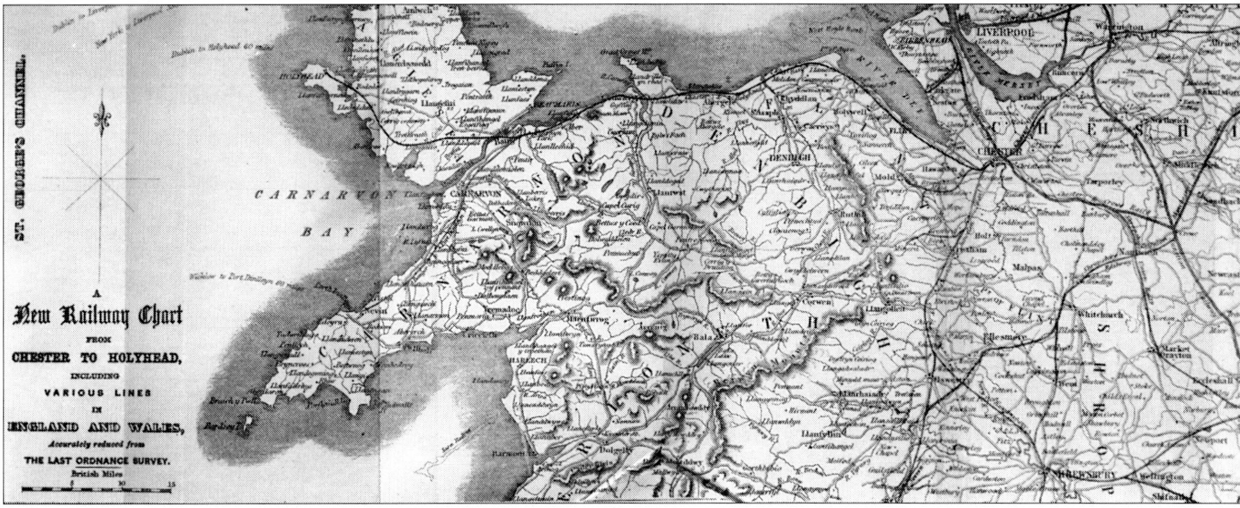

Map of Chester and Holyhead Railway from Parry's Railway Companion from Chester to Holyhead, *1848. Cat. No. 14*

23 William Fairbairn. Correspondence with Robert Stephenson

24 William Fairbairn. An account of the construction of the Britannia and Conway tubular bridges, 1849 (Edwin Clark's personal copy)

25 Sir F. Head. Highways and dry-ways; or the Britannia and Conway tubular bridges, 1861

26 T. Jackson. The tourist's guide to the Britannia Bridge, 1850

27 J. Lucas. Conference of engineers at Britannia Bridge [group portrait]

28 [Anon]. Photograph of Britannia Bridge under construction.

29 T.L. Donaldson. Tubular bridges. Civil engineer and architects journal, 1849

30 Suir Viaduct, Engineer, 1, 1856, 336

31 W. Fairbairn. On tubular girder bridges. Min Procs ICE, 9, 1849-1850, 233-287

32 W. Fairbairn. On the security and hint of strength of tubular girder bridges constructed of wrought iron. Memoirs, Manchester Lit & Phil Soc, 9, 1851, 186

33 J. Hodges. Construction of the Great Victoria Bridge in Canada, 1860

34 [Notman]. Photographs of the Victoria Bridge, Montreal

35 New South Wales Public Works Department. Photograph of tubular girder railway bridges, c.1880

36 W. Swinburne. Diary of visit to Egypt

37 C. Wild. Benha Bridge, Alexander and Cairo Railway. Allgemeine Bauzeitung, 1853

38 Civil engineer and architects journal. Jacquard punching machine, 1848

39 Official description and illustrated catalogue of the Great Exhibition, 1, 1851, p.251

40 K. Cuhlmann. Eisene Brücken. Allgemeine Bauzeitung, 1852

41 F. Kohl. Beschreibung der Göltzsch und Elsterthal Uberbrückung, 1854

42 Molinos et Pronnier. Traité theorique et pratique de la construction des Pons metalliques

43 B. Baker. Long span railway bridges, 1867

44 J. Barton. On the economic distribution of material in the sides of vertical partitions of wrought iron beams. Min Procs ICE, 14, 443-90

45 W. Humber. A complete treatise on cast and wrought iron bridge construction, 1861

46 W. Pole. The life of Sir William Fairbairn, 1877

47 W.J.M. Rankine. A manual of civil engineering.

48 Report of the Commissioners ... into the application of iron to railway structures, 1849

49 S. Smiles. Life of George and Robert Stephenson, 1861

50 R. Stephenson. Iron bridges. Encyclopaedia Britannica, 8th ed.

51 Science Museum. Model of raising the Britannia Bridge tubes. (Courtesy of the Board of Trustees of the Science Museum)

52 Fragment of Britannia Bridge tube

Fabricating the Britannia tubes

The vital statistics

Main spans Two each of 460ft clear.

Side spans Two each of 230ft clear.

Tubes
30ft high at centre, 23ft at abutments.
External width – 14ft 8in
Overall length of completed tube – 1511ft (460.9m).
Weight of each main tube – 1553 tons.
Computed central breaking load – 2537 tons.
Weight of each short tube – 600 tons.
Weight of each completed tube with cast iron stiffeners and permanent way 5270 tons.

Towers
Britannia
55ft x 45ft 5in at deck level, 200ft high above high water, containing 148,625 cu ft of limestone, 144,625 cu ft of sandstone.
Total weight about 20,000 tons.

Shore towers
55ft x 32ft at deck level. 190ft high above high water.

Costs
Experimental work – £3,986.
Towers and abutments – £158,704.
Construction and erection of tubes – £443,161.
TOTAL – £601,865.

Conversion factors
1 inch = 25.4 millimetres.
1 foot = 0.305 metres.
1 ton = 1.016 tonne.
1 ton/ft run = 3.335 tonne/m run.
1 ton/sq in = 15.44N/sq mm.
1 pound/sq in = 0.0069N/sq mm.

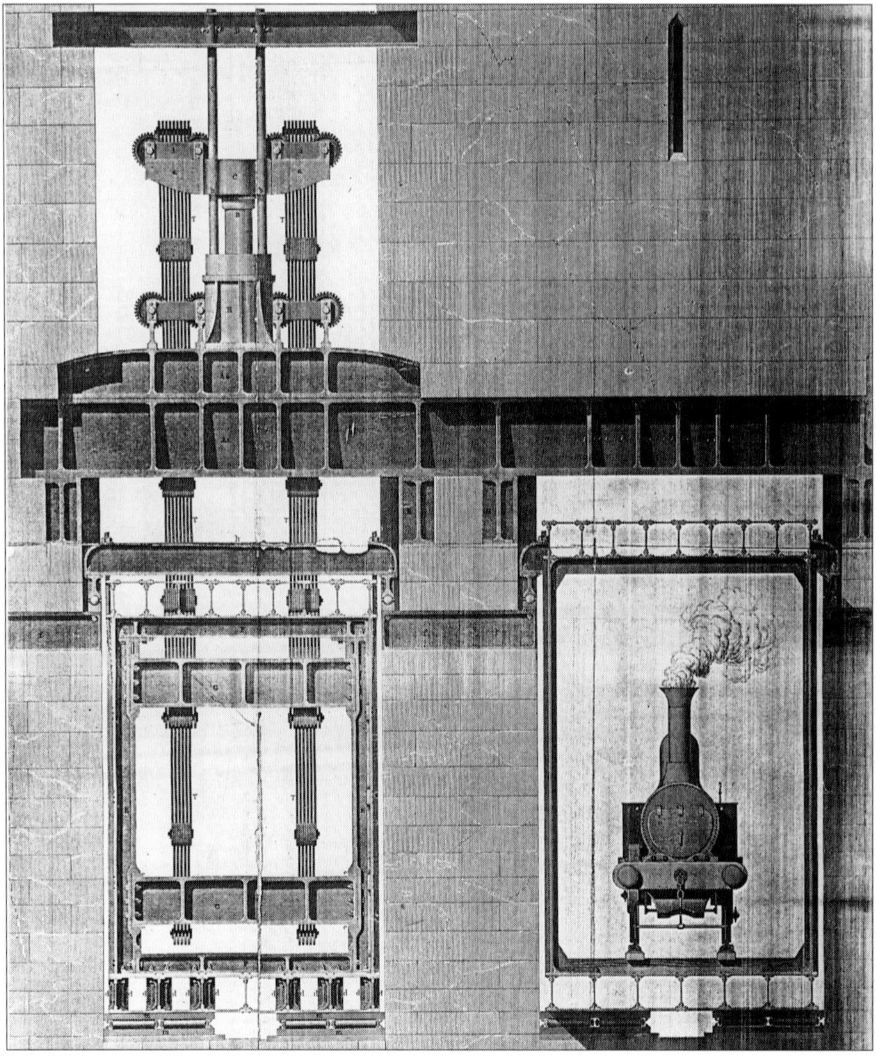

Transverse section of Conwy tube showing lifting apparatus and completed tube

Chronology

14/03/44	Chester & Holyhead Bill.
04/07/44	Royal Assent granted.
By 08/44	Route via Britannia crossing approved by C&HR.
01/03/45	First sod for railway cut.
In 03/45	First drawings for elliptical tubular bridge.
07/04/45	Captain Vidal reports. (On site at Britannia Rock, concerning navigation and minimum channel clearances.)
14/04/45	J.M. Rendel reports.
16/04/45	Sir John Rennie reports.
mid 4/45	Fairbairn visits Stephenson's office.
05/05/45	Chester & Holyhead Bill No.2. House of Commons Committee. Stephenson proposes tubular bridge.
30/06/45	Act receives Royal Assent.
06/07/45	First circular tube test at Millwall.
07/07/45	Second circular test
11/07/45	Third circular test
12/07/45	Fourth & fifth circular tests.
19/07/45	Stephenson invited to Millwall tests on 30/7/45.
21/07/45	Stephenson presses for tests on elliptical tubes.
30/07/45	Sixth, seventh & eighth circular tests.
31/07/45	Ninth circular test.
31/07/45	First, second & third rectangular tube tests.
01/08/45	Fourth & fifth rectangular tests.
02/08/45	Sixth rectangular test
03/08/45	Seventh rectangular test.
04/08/45	Eighth rectangular test.
05/08/45	Ninth rectangular test.
06/08/45	First elliptical tube test.
06/08/45	Visit to Millwall by several Directors and Secretary
10/08/45	Tenth rectangular test.
20/08/45	Eleventh & twelfth rectangular tests.

07/09/45	Second elliptical test.
18/09/45	Third, fourth & fifth elliptical tests.
19/09/45	Eaton Hodgkinson's first visit to Millwall.
19/09/45	Sixth elliptical test.
20/09/45	Thirteenth rectangular test.
20/09/45	Fairbairn suggests square or circular cells for top.
08/10/45	Tenth & eleventh circular tests.
09/10/45	Twelfth circular test.
09/10/45	Fourteenth rectangular test.
14/10/45	Fifteenth rectangular test. 2 circular top cells.
In 10/45	Stephenson goes to Italy for six weeks.
In 11/45	Contract drawings completed, signed by Fairbairn.
09/02/46	Reports submitted to Directors by RS, EH, WF.
11/02/46	C&HR Board approves Thompson's drawings for piers.
Mid 03/46	Hodgkinson continues experiments for Stephenson and Fairbairn continues tube development independently.
17/03/46	Contract drawings completed for piers.
10/04/46	Britannia foundation stone laid by Frank Forster.
24/04/46	Fairbairn orders model tube with cellular top.
12/05/46	Conway foundation stone laid by Alexander M.Ross.
13/05/46	C&HR Board appoints Fairbairn joint engineer for bridges with Stephenson.
In 06/46	Howard & Ravenhill consulted re chains.
11/06/46	Fairbairn completes first drawings for workshops.
In 07/46	Blair completes working drawings for tubes.
08/07/46	Tenders accepted for workshops.
14/07/46	Clark obtains idea for lifting tubes into place.
15/07/46	Fairbairn sends proposals for floating tubes to RS.
29/07/46	Tenders accepted for 8000 tons of iron.
07/08/46	G.B. Airy consulted by Clark.
11/08/46	Sites decided on for staging and workshops.
In 08/46	Decision to erect lifting presses above tubes.

In 08/46	First orders for plates sent out.
mid08/46	Stephenson goes abroad for 5/6 weeks.
Mid 09/46	British Association meeting at Southampton. Hodgkinson lays claim to concept of cellular tops.
27/09/46	Preliminary specifications for tubes posted to potential contractors.
06/10/46	Tenders for tubes due for return.
25/10/46	Stephenson claims concept of cellular bridge at Ware.
06/11/46	Final working drawings completed.
11/01/47	Sites pegged out for staging and workshops.
15/01/47	Drawings for staging and workshops completed.
mid01/47	Directors complain of rising costs of experiments.
25/02/47	Clark suggests single row of top cells.
27/02/47	Hodgkinson's experiments finish.
In 02/47	Staging at Menai for 2 tubes completed.
03/03/47	Stephenson approves single row of top cells.
24/03/47	Fairbairn completes drawings for lifting gear.
08/04/47	First rivet driven at Conway by Clark.
24/04/47	Easton & Amos complete drawings for presses.
12/05/47	Tenders accepted for presses and lifting chains.
13/06/47	First delivery of iron to Menai.
By 07/47	Completion of staging for Menai tubes
10/08/47	First rivet driven at Menai by Edwin Clark.
23/08/47	Stephenson confirms that Fairbairn should equally share all responsibilities on the bridges.
In 11/47	Lifting chains and presses delivered from London.
03/01/48	First Conway tube completed.
In 01/48	I.K. Brunel and Captain Claxton visit Conway.
24/01/48	First tube on supports ready for deflection tests.
20/02/48	Stephenson arrives at Conway to supervise floating.
06/03/48	First Conway tube floated out.

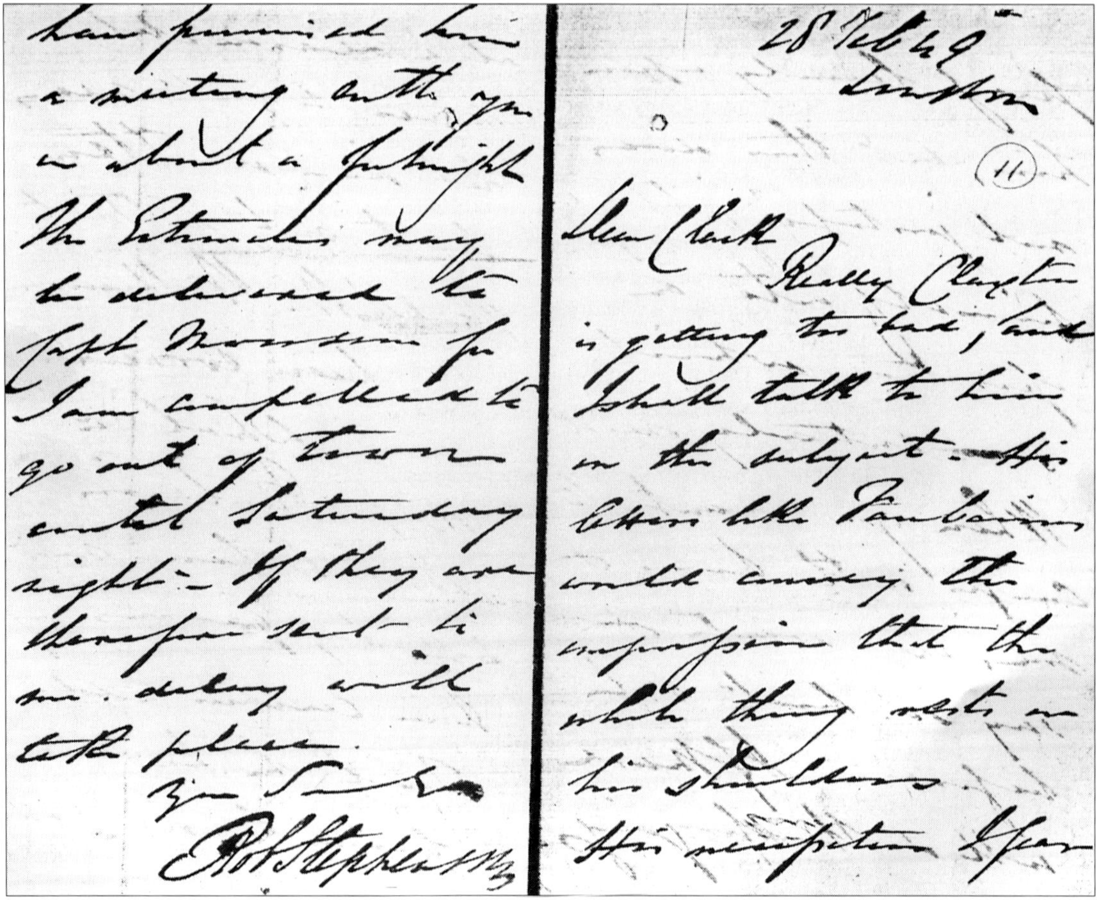

have promised him a meeting with you in about a fortnight. The Estimates may be delivered to Capt Moorsom for I am compelled to go out of town until Saturday night — If they are therefore sent to me no delay will take place.

Yours truly

Robt Stephenson

28 Feb 49
London

Dear Clark

Really Clayton is getting too bad, and I shall talk to him on the subject — His letters like [illegible] would convey the impression that the whole thing rests on his shoulders. His presumption I fear

Letter from Robert Stephenson to Edwin Clark, 28 February 1849

11/03/48	Above ready for lifting.
16/04/48	First Conway tube in final position.
18/04/48	Stephenson drives first locomotive over Conway tube
End 04/48	Captain Simmons inspects first Conway tube.
01/05/48	First Conway tube opened to traffic.
16/05/48	Fairbairn advises RS of intention to resign.
17/05/48	Stephenson's speech at Conway claiming priority for conception of cellular structure at Ware.
22/05/48	Fairbairn submits formal resignation to Directors.
27/07/48	Lines on Anglesey approved by Captain Simmons.
01/08/48	Lines on Anglesey opened for public service.
12/08/48	Death of George Stephenson.
30/08/48	Stephenson's carriage run into at Conway.
12/10/48	Second Conway tube floated out.
In 11/48	Work in hand on first Britannia land tubes.
16/12/48	Second Conway tube ready for traffic.
02/01/49	Captain Simmons inspects second Conway tube.
22/02/49	Shore piers completed.
In 04/49	C&HR Board authorises work on Britannia 2nd line.
04/05/49	First Britannia (Anglesey) tube completed.
20/06/49	First Britannia (Anglesey) tube floated out.
22/06/49	Britannia tower topped out by Stephenson.
10/08/49	First Anglesey tube lifting starts.
17/08/49	Hydraulic press fails. Lifting stopped.
01/10/49	New press. Lifting restarted.
06/10/49	First tube lifting completed.
09/11/49	First Anglesey tube in final position. Anglesey main and shore tubes connected.
20/11/49	Anglesey hydraulic press falls in water.
end11/49	Stephenson arrives at Menai.
04/12/49	Second (Caernarfon) tube floated out.

07/01/50	Second (Caernarfon) tube in place.
03/03/50	Caernarvon shore tube lowered and joined.
04/03/50	Tubes united end to end
05/03/50	Last rivet. First train passes through.
15/03/50	Captain Simmons inspects first completed tube.
18/03/50	Single line opened for passengers.
10/06/50	Third (Anglesey) tube floated out.
11/07/50	Third (Anglesey) tube in final position.
25/07/50	Fourth (Caernarfon) tube floated out.
31/08/50	Stephenson offered knighthood at Newcastle.
19/10/50	Captain Simmons inspects completed Britannia.
??/08/51	Grand banquet for Stephenson at Bangor Ferry.
13/10/52	Queen Victoria and Prince Albert cross Britannia.